AF758493

BIOTECHNOLOGY IN AGRICULTURE, INDUSTRY AND MEDICINE

FOLLOW-ON BIOLOGICS: BACKGROUND AND ISSUES

Biotechnology in Agriculture, Industry and Medicine

Additional books in this series can be found on Nova's website under the Series tab.

Additional E-books in this series can be found on Nova's website under the E-books tab.

Health Care Issues, Costs and Access

Additional books in this series can be found on Nova's website under the Series tab.

Additional E-books in this series can be found on Nova's website under the E-books tab.

BIOTECHNOLOGY IN AGRICULTURE, INDUSTRY AND MEDICINE

FOLLOW-ON BIOLOGICS: BACKGROUND AND ISSUES

DANIEL J. GUTIÉRREZ
EDITOR

Nova Science Publishers, Inc.
New York

Copyright © 2011 by Nova Science Publishers, Inc.

All rights reserved. No part of this book may be reproduced, stored in a retrieval system or transmitted in any form or by any means: electronic, electrostatic, magnetic, tape, mechanical photocopying, recording or otherwise without the written permission of the Publisher.

For permission to use material from this book please contact us:
Telephone 631-231-7269; Fax 631-231-8175
Web Site: http://www.novapublishers.com

NOTICE TO THE READER

The Publisher has taken reasonable care in the preparation of this book, but makes no expressed or implied warranty of any kind and assumes no responsibility for any errors or omissions. No liability is assumed for incidental or consequential damages in connection with or arising out of information contained in this book. The Publisher shall not be liable for any special, consequential, or exemplary damages resulting, in whole or in part, from the readers' use of, or reliance upon, this material. Any parts of this book based on government reports are so indicated and copyright is claimed for those parts to the extent applicable to compilations of such works.

Independent verification should be sought for any data, advice or recommendations contained in this book. In addition, no responsibility is assumed by the publisher for any injury and/or damage to persons or property arising from any methods, products, instructions, ideas or otherwise contained in this publication.

This publication is designed to provide accurate and authoritative information with regard to the subject matter covered herein. It is sold with the clear understanding that the Publisher is not engaged in rendering legal or any other professional services. If legal or any other expert assistance is required, the services of a competent person should be sought. FROM A DECLARATION OF PARTICIPANTS JOINTLY ADOPTED BY A COMMITTEE OF THE AMERICAN BAR ASSOCIATION AND A COMMITTEE OF PUBLISHERS.

Additional color graphics may be available in the e-book version of this book.

Library of Congress Cataloging-in-Publication Data

Follow-on biologics : background and issues / editor, Daniel J. Gutiirrez.
p. ; cm.
Includes bibliographical references and index.
ISBN 978-1-61122-076-6 (hardcover)
1. Pharmaceutical biotechnology. 2. Pharmaceutical biotechnology--Law and legislation--United States. I. Gutiirrez, Daniel J.
[DNLM: 1. Biological Products--standards--United States. 2. Drug Approval--United States. 3. Patents as Topic--United States. QW 800]
RS380.F65 2010
615'.19--dc22
2010041309

Published by Nova Science Publishers, Inc. † New York

Contents

Preface

A biologic is a preparation, such as a drug or a vaccine, that is made from living organisms. A follow-on biologic, or biosimilar, is similar to the brand-name (innovator) product made by the pharmaceutical or biotechnology industry. In contrast to a biologic, most commonly used drugs are synthesized via a chemical process. Biologics often require special handling (such as refrigeration) and are usually administered to patients via injection or infused directly into the bloodstream. This book provides a brief introduction to the relevant law, the regulatory framework at the FDA, the scientific challenges for the FDA in considering the approval of follow-on biologics, and a brief description of the biologics provisions in the new healthcare reform bill signed into law under President Obama, entitled the Patient Protection and Affordable Care Act (PPACA)

Chapter 1- On March 23, 2010, President Obama signed into law a comprehensive health care reform bill, the Patient Protection and Affordable Care Act (PPACA; P.L. 111-148). PPACA establishes a new regulatory authority within the Food and Drug Administration (FDA) by creating a licensure pathway for follow-on biologics, also called biosimilars, and authorizing the agency to collect associated fees.

A biologic is a preparation, such as a drug or a vaccine, that is made from living organisms. A follow-on biologic, or biosimilar, is similar to the brand-name (innovator) product made by the pharmaceutical or biotechnology industry. In contrast to a biologic, most commonly used drugs are synthesized via a chemical process. Biologics often require special handling (such as refrigeration) and are usually administered to patients via injection or infused directly into the bloodstream.

Chapter 2- Congressional interest in the biotechnology industry parallels congressional attention to the availability of lower-cost pharmaceuticals. The market for biologic drugs is expanding by a number of measures including the quantity of approved products, the size of the market, and the importance of these pharmaceuticals to the health of U.S. citizens. Concurrently, patents on many biologics are expected to expire in the next few years. Some commentators have expressed concerns that patent expirations may not be accompanied by the introduction of competing, lower-cost follow-on drugs (also known as biosimilars) in the marketplace.

Chapter 3- Biologics, which are sometimes termed biopharmaceutials or biotechnology drugs, have begun to play an increasingly important role in U.S. health care. Not only are sales of biologics growing rapidly, some experts estimate that in coming years half of all newly approved drugs will result from biotechnology.

A number of patents pertaining to certain biological products will expire in the near future. Some congressional concern has been voiced over the possibility that these patent expirations may not be accompanied by the introduction of competing, lower-cost biologics in the marketplace. With respect to traditional pharmaceuticals, the Drug Price Competition and Patent Term Restoration Act of 1984, a statute commonly known as the "Hatch-Waxman Act," is widely believed to have encouraged the availability of generic substitutes for brand-name pharmaceuticals upon patent expiration. The Act in part permitted the Food and Drug Administration to expedite its marketing approval proceedings with respect to generic drugs.

Chapter 4- Over the past three decades, the lure of patent protection, coupled with the ability to price at market rates, has spurred pioneer drug manufacturers to develop new therapeutic drugs known as biologics. These innovations have improved medical treatments, reduced suffering, and saved the lives of many Americans. Biologic drugs are protein-based and derived from living matter or manufactured in living cells using recombinant DNA biotechnologies. The therapeutic proteins that form the basis of these biologic drugs are far more complex and much larger than the chemically synthesized, small molecules that form the basis of most pharmaceutical products.

In: Follow-On Biologics: Background and Issues
Editor: Daniel J. Gutiérrez
ISBN: 978-1-61122-076-6
© 2011 Nova Science Publishers, Inc.

Chapter 1

FDA Regulation of Follow-On Biologics

Judith A. Johnson

Summary

On March 23, 2010, President Obama signed into law a comprehensive health care reform bill, the Patient Protection and Affordable Care Act (PPACA; P.L. 111-148). PPACA establishes a new regulatory authority within the Food and Drug Administration (FDA) by creating a licensure pathway for follow-on biologics, also called biosimilars, and authorizing the agency to collect associated fees.

A biologic is a preparation, such as a drug or a vaccine, that is made from living organisms. A follow-on biologic, or biosimilar, is similar to the brand-name (innovator) product made by the pharmaceutical or biotechnology industry. In contrast to a biologic, most commonly used drugs are synthesized via a chemical process. Biologics often require special handling (such as refrigeration) and are usually administered to patients via injection or infused directly into the bloodstream.

The new regulatory pathway is analogous to the FDA's authority for approving generic chemical drugs under the Drug Price Competition and Patent Term Restoration Act of 1984 (P.L. 98-4 17). Often referred to as the Hatch-Waxman Act, this law allows the generic company to establish that its drug product is chemically the same as the already approved innovator drug, and thereby relies on the FDA's previous finding of safety and effectiveness for the approved drug. The generic drug industry achieves cost savings by avoiding the expense of clinical trials, as well as the initial drug research and development costs that were incurred by the brand-name manufacturer.

The cost of specialty drug products, such as biologics, is often prohibitively high. For example, the costs per year (in 2009) of some commonly used biologic drugs: Enbrel for rheumatoid arthritis, \$26,000; Herceptin for breast cancer, \$37,000; Rebif for multiple sclerosis, \$40,000; Humira for Crohn's disease, \$51,000; and Cerezyme for Gaucher 's disease, \$200,000. A pathway enabling the FDA approval of follow-on biologics will allow for market competition and reduction in prices, though perhaps not to the same extent as that which occurred with generic chemical drugs under Hatch-Waxman (P.L. 98-4 17).

In contrast to chemical drugs, which are small molecules and for which the equivalence of chemical composition between the generic drug and innovator drug is relatively easy to determine, a biologic, such as a protein, is much larger in size and much more complex in structure. Therefore, comparing a follow-on protein with the brand-name product is more scientifically challenging than comparing chemical drugs. In many cases, current technology will not allow complete characterization of biological products. Additional clinical trials may be necessary before the FDA would approve a follow-on biologic.

This chapter provides a brief introduction to the relevant law, the regulatory framework at the FDA, the scientific challenges for the FDA in considering the approval of follow-on biologics, and a brief description of the biologics provisions in PPACA. Economic studies on potential savings to the federal government over 10 years due to the use of follow-on biologics have ranged between nothing and $14 billion.

On March 23, 2010, President Obama signed into law a comprehensive health care reform bill, the Patient Protection and Affordable Care Act (PPACA; P.L. 111-148). PPACA establishes a new regulatory authority within the Food and Drug Administration (FDA) by creating a licensure pathway for follow-on biologics.[1] This pathway is analogous to that which allowed for the approval of generic chemical drugs via passage of the Drug Price Competition and Patent Term Restoration Act of 1984 (P.L. 98-417), often referred to as the Hatch-Waxman Act.[2] By offering an alternative to brand-name drug products, the Hatch-Waxman Act is credited with lowering the cost of drugs to consumers, as well as allowing for the expansion of the generic drug industry in the United States. This chapter provides a brief introduction to the relevant law, the regulatory framework at the FDA, the scientific challenges for the FDA in considering the approval of follow-on biologics, and a brief description of the biologics provisions in PPACA.[3]

Background

At the time that Hatch-Waxman was being debated by Congress and subsequently implemented by the FDA, the biotechnology industry was just beginning to develop its first biologics for use as human therapeutic agents. The first FDA approval of a biotechnology drug for human use, human insulin, occurred in 1982, followed by human growth hormone in 1985, alpha interferon in 1986, tissue plasminogen activator in 1987, and erythropoietin in 1989. Most biologics require special handling (such as refrigeration) and are usually administered to patients via injection or infused directly into the bloodstream.

Biologic vs. Follow-on Biologic

A *biologic* is a preparation, such as a drug or a vaccine, that is made from living organisms. In contrast, a chemical drug is synthesized via a chemical process. A *follow-on biologic* is similar but not identical to the brand- name, or innovator, product made by the pharmaceutical or biotechnology industry; *biosimilar* is the term used in the European Union.

Biotechnology products are expected to become a larger share of the drugs sold by the pharmaceutical industry to U.S. consumers. However, with no equivalent to the generic alternatives for chemical drugs, the cost of therapeutic biologics is often prohibitively high for individual patients. For example, the costs per year (in 2009) of some commonly used biologic drugs: Enbrel for rheumatoid arthritis, $26,000; Herceptin for breast cancer, $37,000; Rebif for multiple sclerosis, $40,000; Humira for Crohn's disease, $51,000; and Cerezyme for Gaucher's disease, $200,000.[4] Spending by Medicare in 2006 on Epogen, a treatment for anemia, was $2.8 billion, more than the entire FY2006 budget for FDA, which was $1.863 billion.[5] Medicare spending on biologics totaled about $13 billion in 2007.[6] Spending on all pharmaceuticals currently represents about 11% of health care spending in the United States.

Increasingly insurers are adopting strategies to manage the use of expensive biologics, such as a tiered formulary (resulting in higher patient cost-sharing) and prior authorization (requiring physicians to obtain approval from an insurer for coverage before a prescription is filled or administered).[7] In situations in which such expensive drugs have no effect on overall survival, insurers may turn to "innovative payment models, such as providing payment only in cases in which a drug actually leads to clinical improvement."[8]

Market Competition

For chemical drugs, some experts argue that "generic medications decrease prices 60% to 90% on branded oral-solid medications."[9] The Congressional Budget Office estimated the savings generated by generic drug use in 1994 was between $8 billion and $10 billion.[10] The generic drug industry achieves these cost savings by avoiding the expense of clinical trials, as well as the initial drug research and development costs that were incurred by the brand-name manufacturer. In the case of chemical pharmaceuticals, before a generic drug can be marketed, the generic drug company must demonstrate to the FDA that the drug product is identical to the original product. This "sameness" allows the generic company to rely on or "reference" the FDA's previous finding of safety and effectiveness for the approved drug.

Even though patents for several specialty biotechnology drug products have expired, very few have had to face the same type of market competition that occurs with chemical drugs. In contrast to the relatively simple structure and manufacture of chemical drugs, follow-on biological products, with their more complex nature and method of manufacture, will not be identical to the brand-name product, but may instead be shown to be similar. The Generic Pharmaceutical Association (GPhA) advocated that the FDA establish a regulatory system for the approval of follow-on biologics under its existing statutory authority.[11] However, the Biotechnology Industry Organization (BIO) filed a citizen petition with the FDA requesting a number of actions that would inhibit the approval of follow-on biologics.[12]

On April 12, 2006, the European Commission approved the first biosimilar product Omnitrope, a human growth hormone, in Europe following a positive scientific opinion issued by the European Medicines Agency (EMA); a second biosimilar human growth hormone, Valtropin, was approved on April 24, 2006.[13] Sales of Omnitrope in the United States only occurred following the April 10, 2006, ruling by the U.S. District Court in the

District of Columbia in favor of Omnitrope's sponsor, Sandoz. The court ruled that the FDA must move forward with consideration of the abbreviated application, submitted by Sandoz in 2003, that presents Omnitrope as "indistinguishable" from the FDA-approved Genotropin marketed by Pfizer. Sandoz "alleged that the FDA had violated its statutory obligation to act on the Omnitrope application within 180 days, a time frame that the FDA characterized as merely a congressional aspiration."[14]

Federal Trade Commission Report

In June 2009 the Federal Trade Commission (FTC) released a report that examines the potential impact of follow-on biologics on the price of biologic drugs and compares this with the impact of generic drugs on the chemical drug market.[15] The report found that for a number of reasons the competition between follow-on and brand-name biologics is unlikely to be similar to generic and brand-name chemical drug competition.[16] Based on these findings, the FTC report concluded that the 12- to 14-year market exclusivity period is too long to promote innovation, particularly since the brand-name firm will retain substantial market share after the follow-on biologic enters the market. The report also stated that follow-on biologic manufacturers are unlikely to require a 180- day marketing exclusivity period as an incentive to develop interchangeable products. Lastly, the report concluded that special procedures to resolve patent issues are unnecessary and could undermine patent incentives and harm consumers.

The Obama Administration believed that its proposal of a seven-year market exclusivity period was "a generous compromise between what the FTC research has concluded and what the pharmaceutical industry has advocated."[17] In contrast, a group representing the biotechnology industry believed that 12 to 14 years of market exclusivity was necessary to promote innovation and that patent resolution procedures would benefit patients, physicians, insurers, and manufacturers.[18] Others provided commentary and recommendations on the length of the data exclusivity period, including John Calfee of the American Enterprise Institute,[19] Alex Brill, a principal at Matrix Global Advisors, LLC and former chief economist to the House Ways and Means Committee (seven years),[20] and Duke University economist Henry Grabowski (12.9 to 16.2 years).[21]

Economic Studies on Potential Savings

Economic studies on potential savings to the federal government over 10 years due to the use of follow-on biologics have ranged "between nothing and $14 billion."[22] A study by Avalere Health estimated "government savings at $3.6 billion in the first 10 years"; another study by Express Scripts estimated "10-year consumer savings at $71 billion and federal savings at $14 billion."[23] On June 25, 2008, the Congressional Budget Office (CBO) released a cost estimate on a bill introduced in the 110th Congress, S. 1695 (Kennedy). The CBO study found that enactment of S. 1695 would save the federal government $5.9 billion over 10 years (2009-2018) and would reduce total expenditures on biologics in the United States by about $25 billion over the same period.[24] The Obama Administration FY2010 budget proposal estimated the amount of savings, following the implementation of a pathway for the approval of follow-on biologics, at $9.2 billion over a 10-year period (2010-2019).[25]

Patient Safety

Although most observers agree that lower prices for biologics would be of great benefit both to consumers and payers, some have expressed concern that the abbreviated application process allowing for expedited FDA approval of these complex therapeutics might compromise patient safety. A report published in October 2008 investigated the nature, frequency and timing of safety-related regulatory actions for biologics approved in the United States and the European Union.[26] It found that 41 of 174 biologics approved since 1995 were the subject of 82 regulatory actions regarding safety. These regulatory actions were primarily letters to healthcare professionals and some "black box" warnings on product labels but no product withdrawals.

The October 2008 study found that the probability of a first safety-related regulatory action was 14% 3 years after approval and 29% 10 years after approval. The authors noted that these may be underestimates—not all drugs are marketed right after approval (and some may never be marketed), but all biologics that obtained market authorization were included in the study. As is the case with chemically produced drugs, many safety problems are identified only after drug approval because some serious adverse drug effects are rare and will only become apparent following use in large numbers of patients. Lastly, and perhaps most importantly for individuals interested in follow-on biologics products, the study found that the first biologics approved in a chemical, pharmacological and therapeutic subgroup (in other words, innovator products) were especially prone to safety-related regulatory action compared with later approved products in that subgroup.

Relevant Laws

In general, biological products are regulated (*licensed* for marketing) under the Public Health Service Act—originally by the National Institutes of Health (NIH) and its precursors and later, starting in 1972, by the FDA—and chemical drugs are regulated (*approved* for marketing) under the Federal Food Drug and Cosmetic Act—by the FDA. This section provides a brief history of these two Acts and other relevant laws, as well as some of the important amendments that have occurred during the past 100 years.

Biologics Control Act of 1902

The regulation of biologics by the federal government began with the Biologics Control Act of 1902, "the first enduring scheme of national regulation for any pharmaceutical product."[27] The act was groundbreaking, "the very first premarket approval statute in history."[28] It set new precedents, "shifting from retrospective post-market to prospective pre-market government review."[29] The Biologics Act was passed in response to deaths (many in children) from tetanus contamination of smallpox vaccine and diphtheria antitoxin. The act focused on the manufacturing process of such biologic products and required an inspection of the manufacturing facility before a federal license was issued to market the product.

Pure Food and Drugs Act and the Federal Food Drug and Cosmetic Act

The Biologics Act predates the regulation of drugs under the Pure Food and Drugs Act, which was enacted in 1906. The 1906 Act "did not include any form of premarket control over new drugs to ensure their safety ... [and] did not include any controls over manufacturing establishments, unlike the pre-existing Biologics Act and the later-enacted Federal Food Drug and Cosmetic Act (FDC Act)."[30] The Pure Food and Drugs Act was replaced by the FDC Act in 1938. The FDC Act required that drug manufacturers submit a new drug application (NDA) prior to marketing that demonstrated, among other things, that the product was safe.[31]

The Public Health Service Act

The Biologics Act was revised and re-codified (42 USC 262) when the Public Health Service Act (PHS Act) was passed in 1944. The 1944 Act specified that a biological product that has been licensed for marketing by the FDA under the PHS Act is also subject to regulation (though not approval) under the FDC Act. A biological product is defined under section 35 1(i) of the PHS Act, as

> a virus, therapeutic serum, toxin, antitoxin, vaccine, blood, blood component or derivative, allergenic product, or analogous product ... applicable to the prevention, treatment or cure of a disease or condition of human beings.

Section 351(j) of the PHS Act states that "the FDC Act applies to a biological product subject to regulation under this section, except that a product for which a license has been approved under subsection (a) shall not be required to have an approved application under section 505 of such Act." Most biological products regulated under the PHS Act also meet the definition of a drug under section 201(g) of the FDC Act:

> articles intended for use in the diagnosis, cure, mitigation, treatment, or prevention of disease in man or animals; and articles (other than food) intended to affect the structure or any function of the body of man or other animals.

The PHS Act was amended by FDA Modernization Act of 1997 (FDAMA) to require a single biological license application (BLA) for a biological product, rather than the two licenses— Establishment License Application (ELA) and Product License Application (PLA)—that had been required between 1944 and 1997. The PHS Act provides authority to suspend a license immediately if there is a danger to public health.

Natural Source Biological Products and the Hatch-Waxman Act

As stated previously, biological products are, in general, regulated—licensed for marketing— under the PHS Act, and chemical drugs are regulated—approved for marketing—under the FDC Act. However, through a historical quirk, the FDA was given regulatory authority over certain natural source biological products; these products have been regulated as drugs under the FDC Act rather than as biologics under the PHS Act. In 1941, three years prior to the re-codification of the Biologics Act, Congress gave the FDA authority over the marketing of insulin.[32] Insulin is a peptide hormone, a small protein that regulates carbohydrate metabolism.[33] In the 1940s, insulin "was obtained in the same manner as many biologics, namely extraction from animals. Despite this similarity with biologics, insulin was regulated by FDA."[34] In addition to insulin, the distinction of a biological product regulated as a drug under the FDC Act rather than as a biologic under the PHS Act holds true for a small set of products that are mostly hormones: glucagon, human growth hormone, hormones to treat infertility, hormones used to manage menopause and osteoporosis, and certain medical enzymes (hyaluronidase and urokinase).[35]

This distinction is important because the Hatch-Waxman Act provides a mechanism for the approval of generic drugs under the FDC Act but not under the PHS Act. Generic pharmaceutical companies could seek to gain approval of follow-on biological products for the small number of biologics approved under the FDA Act but not the much larger group of biologics approved under the PHS Act. Hatch-Waxman added two abbreviated pathways to the FDC Act for subsequent versions of already approved products: section 505(j) and section 505(b)(2).

Section 505(j) established an Abbreviated New Drug Application (ANDA) process for a generic drug that contains the *same* active ingredient as the brand-name innovator drug. In the ANDA, the generic company establishes that its drug product is chemically the same as the already approved innovator drug, and thereby relies on the FDA's previous finding of safety and effectiveness for the approved drug. The 505(j) pathway is used for the approval of most generic chemical drugs.

Under the second pathway, a drug that has a significant difference from an innovator drug, but is still sufficiently *similar* to that drug, may be the subject of a 505(b)(2) application. The company filing the application must submit additional non-clinical and clinical data to show that the proposed product is safe and effective.[36] However, the application may rely on published literature or on the FDA's finding of safety and effectiveness for the already approved product to support the approval of the proposed product. The 505(b)(2) pathway has been used to approve Omnitrope, a follow-on human growth hormone, and a few other follow-on protein products.[37] All have been members of the small set of biologic products that were regulated as drugs.

Regulatory Framework

Following enactment of the 1902 Biologics Act, regulatory responsibility for biologics was first delegated to the Hygienic Laboratory, a precursor of NIH.[38] In 1972, regulatory authority for biologics was transferred from the NIH Division of Biological Standards to the FDA Bureau of Biologics, which eventually became the agency's Center for Biologics Evaluation and Research (CBER).[39]

Because biotechnology products frequently cross the conventional boundaries between biologics, drugs, and devices, determining the jurisdictional status of these new products has been difficult for both the FDA and industry. Some products have had characteristics that met multiple statutory and scientific definitions.[40] In 1991, the FDA published an Intercenter

Agreement between CBER and the Center for Drug Evaluation and Research (CDER).[41] In general, the agreement stated that traditional biologics (vaccines, blood, blood products, antitoxins, allergenic products), as well as most biotechnology products, would be regulated by CBER. The small set of biologics mentioned earlier that are regulated as drugs under the FDC Act would continue to be regulated by CDER, regardless of the method of manufacture.

In 2002, however, the FDA announced its intention to reorganize review responsibilities, consolidating review of new pharmaceutical products under CDER, thereby allowing CBER to concentrate on vaccines, blood safety, gene therapy, and tissue transplantation.[42] On June 30, 2003, responsibility for most therapeutic biologics was transferred from CBER to CDER.[43] Under the new structure, biological products transferred to CDER will continue to be regulated as licensed biologics under section 351 of the PHS Act. Examples of products transferred to CDER include monoclonal antibodies; proteins intended for therapeutic use (interferons, thrombolytic enzymes); immunomodulators (other than vaccines and allergenic products); and, growth factors, cytokines, and monoclonal antibodies intended to alter production of blood cells.[44] Remaining at CBER are traditional biologics such as vaccines, allergenic products, antitoxins, antivenins, venoms, and blood and blood products, including recombinant versions of plasma derivatives (clotting factors produced via biotechnology).

As stated previously, the Hatch-Waxman Act added two abbreviated pathways under the FDC Act—505(j) and 505(b)(2)—but not under the PHS Act, for the approval of additional products subsequent to the innovator product. Because of the complex nature of most biological products and their methods of manufacture, such products will not be identical to the brand-name product; therefore, the 505(j) pathway cannot be used for product approval. However, if a biological product is sufficiently similar to the innovator product, the 505(b)(2) pathway may be used by a company for the approval of its biologic. Following the enactment of Hatch-Waxman, the FDA published in 1999 a draft guidance on applications covered by section 505(b)(2); the guidance has never been finalized.[45]

As things currently stand, and as discussed above, the 505(b)(2) pathway has been used only for those biologics that have been regulated as drugs under the FDC Act. However, the vast majority of biologics have been regulated under the PHS Act. The FDA's position is that additional legislation is required to provide such a pathway under the PHS Act. For traditional biologics regulated under the PHS Act, the agency's longstanding policy has been that a full BLA, including clinical testing, would be required for the licensing of each such product. In a 1974 *Federal Register* notice, the FDA stated that

> [u]nlike the regulation of human and animal drugs, all biological products are required to undergo clinical testing in order to demonstrate safety, purity, potency and effectiveness prior to licensing, regardless whether other versions of the same product are already marketed or standards for the product have been adopted by rulemaking. Indeed, many of the existing standards require specific clinical testing before approval will be granted. This is required because all biological products are to some extent different and thus each must be separately proved safe, pure, potent, and effective.... There is no such thing as a "me-too" biologic.[46]

When publishing the final rule on the ANDA procedure that had been outlined in Hatch-Waxman, the FDA stated in 1992 that "these procedures are inapplicable to ... biological drug products licensed under 42 USC 262 (section 351 of the PHS Act)."[47] Most recently, during hearing testimony on May 2, 2007, before the Subcommittee on Health of the House Energy and Commerce Committee, Janet Woodcock, Deputy Commissioner and Chief Medical Officer of the FDA, stated in response to questioning that there is no pathway under the PHS Act for the approval or licensing of follow-on biologics that is similar to the 505(b)(2) pathway under the FDC Act, and that the FDA would be willing to work with Congress in crafting a legislative approach to creating such a pathway.

Scientific Challenges

Comparing a follow-on protein with the brand-name product is more scientifically challenging than comparing generic and brand-name chemical drugs. For chemically synthesized drugs, which are relatively small molecules, the equivalence of chemical composition between the generic drug and innovator drug is relatively easy to determine. In contrast, therapeutic proteins are much larger in size (100- to 1,000-fold larger than chemically synthesized drugs), have a much more complex three-dimensional structure, and may consist of mixtures rather than one pure entity.

A protein is a large organic molecule composed of a long chain of component parts, called amino acids, which are linked by chemical bonds. This amino acid chain folds into a complex three- dimensional structure. Slight changes in the chain or three-dimensional shape can influence the protein's biological activity. All manufactured biologics (brand or follow-on) can vary slightly from lot to lot, even when the manufacturing process has not been changed. Proteins can also be altered by the addition of other chemicals, such as sugar groups (glycosylation), at various points along the amino acid chain. In many cases, current technology will not allow complete characterization of biological products. In prepared testimony before Congress, FDA Deputy Commissioner Janet Woodcock outlined the scientific challenges involved in determining the safety and effectiveness of follow-on biologics, often referred to by FDA as follow-on protein products:

> Current technologies, such as peptide mapping, protein sequencing, and mass spectroscopy enable manufacturers to determine, with certainty, the amino acid sequence of a recombinant protein. However, the amino acid sequence is the most rudimentary characteristic of a protein. Conclusive analysis of other aspects of a protein's structure requires much more sophisticated technologies and is fraught with uncertainties that are proportional to the size and complexity of the protein itself. Such complexities include folding of the protein's amino acid chain into highly organized structures, post-translational modification of the protein with a broad range of biochemical additions (e.g., glycosylation, acetylation, phosphorylation, etc.), and association of multiple protein molecules into aggregates. It is the combination of the protein's amino acid sequence and its structural modifications that give a protein its unique functional characteristics. Therefore, the ability to predict the clinical comparability of two products depends on our understanding of the relationship between the structural characteristics of the protein and its function, as well as on our ability to demonstrate structural similarity between the follow-on protein and the reference product. Although this currently may be possible for some relatively simple protein products, technology is not yet sufficiently advanced to allow this type of comparison for more complex protein products.[48]

Another challenge for the FDA is determining whether the follow-on biologic is sufficiently similar to the brand-name biologic that the two products are interchangeable.

Several terms are important to this discussion. Products that are considered to be *therapeutically equivalent* "are approved drug products, usually made by different manufacturers, that are pharmaceutical equivalents and for which bioequivalence has been demonstrated. Therapeutic equivalents can be expected to have the same clinical effect and safety profile when administered to patients under the conditions specified in the labeling."[49] *Pharmaceutical equivalents* are products that contain the same active ingredient in the same strength, dosage form, and route of administration.[50] *Bioequivalence* means that the products are absorbed into the body at a similar rate and extent.[51] *Interchangeability* "is not defined by FDA and could have a number of different meanings. It could refer to products that are therapeutic equivalents, and thus could, in some circumstances, be substituted at the pharmacy level without a physician's intervention. Alternatively, the term could describe similar products that are not 'substitutable' but which, under a physician's supervision, could be used to treat the same disease or condition in the same patient."[52]

Most drugs approved under section 505(j) are therapeutically equivalent to the already approved drug product. In her testimony, Dr. Woodcock explains the importance of a determination of therapeutic equivalence for a generic drug and the reasons why such a determination for a follow- on protein product may not be possible, at least at the present time:

> In many jurisdictions, therapeutically equivalent drugs may be substituted at the pharmacy level, without a physician's intervention.... Because of the variability and complexity of protein molecules, current limitations of analytical methods, and the difficulties in manufacturing a consistent product, it is unlikely that, for most proteins, a manufacturer of a follow-on protein product could demonstrate that its product is identical to an already approved product. Therefore, the section 505(j) generic drug approval pathway, which is predicated on a finding of the same active ingredient, will not ordinarily be available for protein products.[53]

Immunogenicity, or the ability to elicit an immune response, is another important term in the discussion of follow-on proteins. An immune response to a therapeutic protein can range from detectable, but clinically insignificant, to one that can cause safety problems for the patient or limit the effectiveness of the product. For some biologics, such as vaccines, stimulating an immune response is the intended outcome. However, for other types of therapeutic products, an immune response can lower the clinical effect of a protein. Dr. Woodcock describes the implications at length in the prepared testimony:

> Adverse safety events from an immune response could include hypersensitivity reactions such as anaphylaxis, rash, fever and kidney problems, to cross-reaction with an endogenous (naturally occurring in the body) protein (e.g., erythropoietin). Immunogenicity may be influenced by patient-related, disease-related, or product-related factors. Immune responses to administered protein products can be extremely serious or life-threatening; therefore, this issue requires significant attention. The ability to predict immunogenicity of a protein product, particularly the more complex proteins, is extremely limited. Therefore, some degree of clinical assessment of a new product's immunogenic potential will ordinarily be needed. The extent of independent testing needed will again depend on a variety of scientific factors such as the indication, whether the product is to be administered chronically, the overall assessment

of the product's immunogenic potential, and whether there is the possibility of generating a cross-reaction with an important endogenous molecule.

Even if a follow-on protein product is found to be safe and effective by the FDA, this finding does not mean that the follow-on protein product would be interchangeable with, or substitutable for, the originally approved brand-name product. To establish that the follow-on protein product is substitutable for the brand-name product, the manufacturer of the follow-on product must demonstrate through additional clinical data that repeated switches from the follow-on product to the brand-name product (and vice versa) would have no negative effect on the safety and/or effectiveness of the products. In other words, there must be no problems with immunogenicity. "For many follow-on protein products, and, in particular, the more complex proteins, there is a significant potential for repeated switches between products to have a negative impact on the safety and/or effectiveness. Therefore, the ability to make determinations of substitutability for follow-on protein products may be limited."[54]

Legislation

Legislative History

Efforts by Congress to address the need for an pathway for the licensure of follow-on biologics began with the introduction of the Access to Life-Saving Medicine Act (H.R. 6257, Waxman, and S. 4016, Schumer) which was introduced near the end of the 109th Congress and received no further action.

Early in the first session of the 110th Congress, two competing legislative approaches were introduced that would have allowed FDA to approve follow-on biologic products. In general, the approach taken by H.R. 1038 (Waxman) and its companion bill S. 623 (Schumer) were favored by the generic drug industry, while H.R. 1956 (Inslee), and S. 1505 (Gregg) were favored by companies that have developed the innovator or brand-name products. The introduction of S. 1695 (Kennedy) in June 2007 provided a third approach. The bipartisan Senate sponsors (Senators Kennedy, Hatch, Enzi, and Clinton) of S. 1695 claimed to have negotiated a compromise between the brand-name manufacturers and the generic drug industry. The sponsors of S. 1695 intended to add the bill's language to legislation (H.R. 2900/S. 1082/H.R. 3580) reauthorizing FDA's user fee programs when this larger FDA bill entered conference negotiations. However, when H.R. 3580 became P.L. 110-85 on September 27, 2007, it did not include S. 1695. The introduction of H.R. 5629 (Eshoo) in March 2008 provided yet another approach. H.R. 5629 was similar in some respects to S. 1695 with a few important differences that were favored by the brand-name industry. S. 1695 was reported on November 19, 2008. No further action was taken on this or the other follow-on biologics bills in the 110th Congress.

In the 111th Congress, H.R. 1427 (Waxman) contained a number of changes when compared with H.R. 1038 (Waxman) from the 110th Congress.[55] H.R. 1548 (Eshoo) was essentially the same as H.R. 5629 (Eshoo) as introduced in the 110th Congress with a few minor changes. Health care reform legislation also provided a pathway for the approval of follow-on biologics. The biologics provisions were added to PPACA in "Title VII—

Improving Access to Innovative Medical Therapies" as "Subtitle A—Biologics Price Competition and Innovation," sections 7001 through 7003. The following paragraphs briefly describe the biologics provisions in the Patient Protection and Affordable Care Act (PPACA; P.L. 111-148).

Biologics Provisions in PPACA, P.L. 111-148

Interchangeability

A determination on interchangeability is required if the application shows that the biological product (1) is biosimilar to the reference product, (2) can be expected to produce the same clinical result in any given patient, and (3) can be alternated or switched with use of the reference product without risk to the patient in terms of safety or diminished efficacy compared with use of the reference product alone. There is no requirement on the publication of guidance.

Clinical studies

PPACA requires information in the FDA application demonstrating that the follow-on biologic is similar to the reference product based on data from a clinical study or studies to demonstrate safety, purity, and potency for one or more appropriate conditions of use for which the reference product is licensed. However, the Secretary may determine that elements in the application, such as clinical studies, may be unnecessary.

FDA Guidance documents

FDA may publish proposed guidance for public comment prior to publication of final guidance, and if so, FDA must establish a process to allow public input regarding priorities for issuing guidance. The issuance or non-issuance of guidance would not preclude the review of, or action on, an application.

Data exclusivity for reference product

Data exclusivity is separate from the protection provided by a patent. Data exclusivity is the period of time during which the manufacturer of a follow-on product, in the preparation of its application for FDA approval, is blocked from referring to the data submitted in the original application to FDA for the approval of the brand-name product; this results in a period of exclusive marketing for the brand-name product.

PPACA provides a 12-year exclusivity period from the date on which the reference product was first approved. If a reference product has been designated an orphan drug, an application for a biosimilar or interchangeable product may not be filed until the later of (1) the seven-year period of orphan drug exclusivity described in the FFDCA or (2) the 12-year period established by the Senate bill.

Market exclusivity for the first interchangeable product

PPACA allows for a period of exclusive marketing for the follow-on biologic product that is the first to be established as interchangeable with the reference product.

Pa tents[56]

PPACA sets forth provisions governing patent infringement claims against an applicant or prospective applicant for a follow-on biological product license. It establishs new processes for identifying patents that might be disputed between the brand-name company and the company submitting a biosimilar application and also would establish a multistep patent resolution process.

Biological products approved under FFDCA

PPACA stipulates that all biological product applications must be submitted under section 351 of the PHS Act. For the small number of biological products that have been approved under section 505 of the FFDCA, the approved application is deemed to be a license for the biological product under section 351 as of 10 years after enactment.

User fees

The Secretary of HHS must develop recommendations regarding goals for the review of follow- on biologic product applications for FY20 13 through the end of FY20 17 and present them to Congress. The recommendations must be published in the *Federal Register* with a 30-day public comment period, and a public meeting must be held. The revised recommendations would be presented to Congress by January 15, 2012. Based on these recommendations, it is the sense of the Senate that Congress should authorize a user fee program effective October 1, 2012. Through October 1, 2010, the Secretary must collect data on the cost of follow-on biologic application review as conducted according to the prescription drug user fee program. Two years after receiving the first user fee for a follow-on biological product application and every two years thereafter until October 1, 2013, the Secretary must perform an audit of the application review costs. An alteration of the user fee would occur depending on results of the audit, as specified in the bill.

Pediatric biologics[57]

PPACA provides for an extra six months of market exclusivity for a new biologic drug if pediatric studies are conducted prior to FDA approval of the drug. An extra six months of market exclusivity is provided for a biologic drug already on the market if pediatric studies are conducted and the request for the extension is made not less than nine months before the expiration of the original exclusivity period.

PPACA requires an IOM study to be conducted that will review and assess the number and importance of biological products for children that are being tested as a result of these biologics provisions, as well as biological products that are not being tested for pediatric use, and offer recommendations for ensuring pediatric testing of biological products.

Savings

PPACA requires the HHS Secretary and the Treasury Secretary to determine for each fiscal year the amount saved to the federal government as a result of enactment of the approval pathway for biosimilar biological products; the savings will be used for deficit reduction.

End Notes

[1] Also referred to as biosimilars, or sometimes as biogenerics or generic biologics. The FDA and many others consider the use of the word *generic* to be inaccurate because the term has been used, in the context of chemical drugs, to mean identical. The FDA often uses the term *follow-on protein product*, because many biologics are proteins; biosimilar is used in the European Union.

[2] CRS Report R41114, *The Hatch-Waxman Act: A Quarter Century Later*, by Wendy H. Schacht and John R. Thomas.

[3] For patent issues, see CRS Report RL33901, *Follow-On Biologics: Intellectual Property and Innovation Issues*, by Wendy H. Schacht and John R. Thomas.

[4] Alfred B. Engelberg, Aaron S. Kesselheim, and Jerry Avorn, "Balancing Innovation, Access, and Profits--Market Exclusivity for Biologics," *New England Journal of Medicine*, vol. 361, no. 20 (November 12, 2009), pp. 1917-1919.

[5] Personal communication with Lisa Yen, Centers for Medicare and Medicaid Services Office of Legislation, February 13, 2009. FDA's budget is composed of $1.494 billion in direct appropriations and $369 million in user fees; see CRS Report RL34334, *The Food and Drug Administration: Budget and Statutory History, FY1980-FY2007*, coordinated by Judith A. Johnson.

[6] Medicare Payment Advisory Commission, *Report to the Congress: Improving Incentives in the Medicare Program*, Chapter 5: Medicare payment systems and follow-on biologics, Washington, DC, June 2009, pp. 103-133, http://www.medpac.gov/documents/Jun09_EntireReport.pdf.

[7] Thomas H. Lee and Ezekiel J. Emanuel, "Tier 4 Drugs and the Fraying of the Social Compact," *The New England Journal of Medicine*, vol. 359, no. 4 (July 24, 2008), pp. 333-335.

[8] Ibid.

[9] Jonah Houts, testimony before the House Committee on Oversight and Government Reform, March 26, 2007.

[10] Congressional Budget Office, "How Increased Competition from Generic Drugs Has Affected Prices and Returns in the Pharmaceutical Industry," July 1998.

[11] Bill Nixon, President and CEO, Generic Pharmaceutical Association, letter to Daniel Troy, Chief Counsel, FDA, January 18, 2002, at http://www.fda.gov/cder/ogd/GPHA_jan_21.htm.

[12] BIO Citizen Petition, Follow-on Therapeutic Proteins, April 23, 2003, at http://www.fda.gov/OHRMS/ DOCKETS/ DOCKETS/03p0176/03p-0176-cp00001-01-vol1 1 .pdf.

[13] "Europe approves two follow-on human growth hormones," *Nature Biotechnology*, vol. 24 (June 2006), p. 601.

[14] Ibid.

[15] Federal Trade Commission, *Emerging Health Issues: Follow-on Biologic Drug Competition*, Washington, DC, June 10, 2009, http://www.ftc.gov/os/2009/06/P083901biologicsreport.pdf.

[16] The FTC report lists the following six reasons: (1) The substantial costs to obtain FDA approval, plus the substantial costs to develop manufacturing capacity, will limit the number of FOB competitors; (2) The lack of automatic substitution between an FOB drug and a pioneer biologic drug will slow the rate at which FOBs can acquire market share; (3) An FOB drug also may have difficulty gaining market share due to concerns about safety and efficacy differences with the pioneer biologic drug; (4) Biologic drugs currently are not reimbursed according to strategies that insurers often use to encourage the use of lower-priced drugs; (5) As a result of these factors, FOB entry, although important, will be less-dramatic than generic drug competition. FOB entry is likely in biologic drug markets larger than $250 million in annual sales. Only two or three FOB manufacturers are likely to attempt entry for a given pioneer drug product. These entrants are unlikely to introduce their drugs at discounts any larger than between 10 and 30% of the pioneer product's price; (6) The effect on pioneer manufacturers also will be different. They are expected to respond and offer competitive discounts to maintain market share and are likely to retain 70 to 90% of their market share and will continue to reap substantial profits, even after FOB entry.

[17] Letter from Nancy-Ann DeParle, Director, Office of Health Reform, and Peter Orszag, Director, Office of Management and Budget, to Henry A. Waxman, Chairman of the House Energy and Commerce Committee, June 24, 2009, http://energycommerce.house.gov/Press_111/20090625/biologicsresponse.pdf

[18] A rebuttal to the FTC report prepared by the Biotechnology Industry Organization (BIO) can be found at http://bio.org/healthcare/followonbkg/FTC_biosimilars_report_rebuttal.pdf

[19] John Calfee, *When Patents Are Not Enough: Data Exclusivity for Follow-On Biologics*, American Enterprise Institute, Washington, DC, December 2008, http://www.aei.org/docLib/20081208_ 23731 HPO10 Calfee _g.pdf.

[20] Alex M. Brill, *Proper Duration of Data Exclusivity for Generic Biologics: A Critique*, November 2008, http://www.tevadc.com/Brill_Exclusivity_in_Biogenerics.pdf.

[21] Letter from Henry Grabowski, Professor of Economics and Director of the Program in Pharmaceuticals and Health Economics, to Federal Trade Commission, Office of the Secretary, July 6, 2009, http://econ.duke.edu/~grabow/FDS/ FullFTC_response.pdf. See also Henry Grabowski, Iain Cockburn, and Genia Long, et al., "Data Exclusivity Periods and Next Generation Improvements to Innovator Biologics: Key

Issues," Duke University Department of Economics Working Paper, No. 2009-05, April 29, 2009, http://www.econ.duke.edu/Papers/PDF/DWPaper2009-05.pdf.

[22] "CBO Weighs 2 Studies That Show Little Savings From Biogenerics," Inside Health Policy, July 19, 2007.

[23] Ibid.

[24] Congressional Budget Office, Cost Estimate, S. 1695, Biologics Price Competition and Innovation Act of 2007, June 25, 2008, at http://www.cbo.gov/ftpdocs/94xx/doc9496/s1695.pdf.

[25] House Budget Committee, Summary of the President's Fiscal Year 2010 Budget, p. 6 at http://budget.house.gov/pres_budgets/fy2010/02.27.2009_FY2010_Pres_Budget_Summary.pdf.

[26] Thijs J. Giezen, Aukje K. Mantel-Teeuwisse, and Sabine M. J. M. Straus, et al., "Safety-related regulatory actions for biologicals approved in the United States and the European Union," *Journal of the American Medical Association*, vol. 300, no. 16 (October 22, 2008), pp. 1887-1896.

[27] David M. Dudzinski, "Reflections on Historical, Scientific, and Legal Issues Relevant to Designing Approval Pathways for Generic Versions of Recombinant Protein-Based Therapeutics and Monoclonal Antibodies," *Food and Drug Law Journal*, vol. 60, pp. 143-260.

[28] Ibid, p. 147.

[29] Ibid.

[30] Gary E. Gamerman, "Regulation of Biologics Manufacturing: Questioning the Premise," *Food and Drug Law Journal*, vol. 49, 1994, pp. 213-235.

[31] For further information, see CRS Report RL32797, *Drug Safety and Effectiveness: Issues and Action Options After FDA Approval*, by Susan Thaul.

[32] Dudzinski, *Food and Drug Law Journal*, vol. 60, p. 153. The Insulin Amendments P.L. 77-366, codified at 21 USC 356, were repealed by P.L. 105-115, the Food and Drug Administration Modernization Act (FDAMA).

[33] A protein is a large organic molecule composed of a long chain or chains of amino acids linked by chemical bonds. Insulin is a short chain of 51 amino acids. Examples of carbohydrates include sugars and starch.

[34] Dudzinski, *Food and Drug Law Journal*, vol. 60, p. 154.

[35] Janet Woodcock, Deputy Commissioner, Chief Medical Officer, FDA, testimony before the House Committee on Oversight and Government Reform, March 26, 2007, at http://oversight 22106.pdf; BIO Citizen Petition, Follow-on Therapeutic Proteins, April 23, 2003, at http://www.fda.gov/OHRMS/ DOCKETS/DOCKETS/ 03p0176/03p-0176-cp00001-01-voll 1 .pdf.

[36] Janet Woodcock, testimony before the House Committee on Oversight and Government Reform, March 26, 2007.

[37] These products are GlucaGen (glucagon recombinant for injection), Hylenex (hyaluronidase recombinant human), Hydase and Amphadase (hyaluronidase), and Fortical (calcitonin salmon recombinant) Nasal Spray. Center for Drug Evaluation and Research, U.S. Food and Drug Administration, *Omnitrope (somatropin [rDNA origin]) Questions and Answers*, May 30, 2006, at http://www.fda.gov/cder/drug/infopage/somatropin/qa.htm.

[38] Ibid., p. 148, and The NIH Almanac—Historical Data: Chronology of Events, at http://www.nih.gov/about/ almanac/ historical/chronology _of _events.htm. In 1937, the biologics control program was assigned to the newly established Division of Biologics Control. In 1955, the biologics control function was placed in the newly formed Division of Biologics Standards.

[39] The NIH Almanac; Donna Hamilton, "A Brief History of the Center for Drug Evaluation and Research," FDA History Office, November 1997, at http://www.fda.gov/cder/about/history/Histext.htm. During the early 1980s, the Bureau of Drugs and the Bureau of Biologics merged to form the National Center for Drugs and Biologics. In 1984, all of the National Centers within FDA were redesignated simply as Centers. In 1987, the Center for Drugs and Biologics was split into the Center for Drug Evaluation and Research (CDER) and the Center for Biologics Evaluation and Research (CBER). CBER continues to use NIH facilities and buildings until the expected move in 2012 to the new FDA headquarters in White Oak, Maryland (see http://www.fda.gov/oc/whiteoak/projectschedule.html).

[40] See, for example, "Assignment of Agency Component for Review of Premarket Applications," Final Rule, Federal Register, vol. 56, no. 225, November 21, 1991, pp. 58754-58758, at http://www.fda.gov/OHRMS/DOCKETS/98fr/91- 27869.pdf.

[41] The Intercenter Agreement is available at http://www.fda.gov/oc/ombudsman/drug-bio.htm.

[42] FDA Press Release, "FDA to Consolidate Review Responsibilities for New Pharmaceutical Products," September 6, 2002, at http://www.fda.gov/bbs/topics/NEWS/2002/NEW00834.html.

[43] *Federal Register*, vol. 68, no. 123, June 26, 2003, pp. 38067-38068.

[44] Transfer of Therapeutic Products to the Center for Drug Evaluation and Research, at http://www.fda.gov/cber/transfer/transfer.htm. Also of interest is Approved Products Transferring to CDER, at http://www.fda.gov/cber/transfer/ transfprods.htm, and Therapeutic Biological Products, at http://www.fda.gov/cder/biologics/default.htm.

[45] Guidance for Industry, Applications Covered by Section 505(b) (2), October 1999, at http://www. fda.gov/CDER/GUIDANCE/2853dft.pdf.

[46] *Federal Register*, v. 39, no. 248, December 24, 1974, p. 44641.

[47] *Federal Register*, v. 57, no. 82, April 28, 1992, p. 17951.

[48] Janet Woodcock, Deputy Commissioner, Chief Medical Officer, FDA, testimony before the Subcommittee on Health, Committee on Energy and Commerce, May 2, 2007, at http://energycommerce.house.gov /cmte_mtgs/110-hehrg.050207.Woodcock-testimony.pdf.

[49] Ibid.

[50] Janet Woodcock et al., "The FDA's Assessment of Follow-on Protein Products: A Historical Perspective," *Nature Reviews Drug Discovery*, published online April 13, 2007, at http://www.nature.com/reviews/drugdisc.

[51] Ibid.

[52] Janet Woodcock, testimony before the Subcommittee on Health, Committee on Energy and Commerce, May 2, 2007.

[53] Ibid.

[54] Ibid.

[55] A companion bill, S. 726 (Schumer), was introduced on March 26, 2009.

[56] This section was written by John Thomas.

[57] This section was written by John Thomas.

In: Follow-On Biologics: Background and Issues
Editor: Daniel J. Gutiérrez

ISBN: 978-1-61122-076-6
© 2011 Nova Science Publishers, Inc.

Chapter 2

P.L. 111-148: Intellectual Property Provisions for Follow-On Biologics

Wendy H. Schacht and John R. Thomas

Summary

Congressional interest in the biotechnology industry parallels congressional attention to the availability of lower-cost pharmaceuticals. The market for biologic drugs is expanding by a number of measures including the quantity of approved products, the size of the market, and the importance of these pharmaceuticals to the health of U.S. citizens. Concurrently, patents on many biologics are expected to expire in the next few years. Some commentators have expressed concerns that patent expirations may not be accompanied by the introduction of competing, lower-cost follow-on drugs (also known as biosimilars) in the marketplace.

Biologics differ from traditional pharmaceuticals in their complexity and mode of manufacture. Typical pharmaceutical products consist of small molecules, on the order of dozens of atoms, that may be readily characterized and reproduced through well-understood chemical processes. In contrast, biologics are often made up of millions of atoms, feature a more intricate structure than traditional pharmaceuticals, and are manufactured from living cells through biological processes.

An expedited approval process and a patent dispute resolution procedure for traditional, small molecule pharmaceuticals was created by the Drug Price Competition and Patent Term Restoration Act of 1984, a statute commonly known as the "Hatch-Waxman Act." This law, which amended the Federal Food, Drug, and Cosmetic Act, is widely believed to have encouraged the availability of generic substitutes for many brand-name pharmaceuticals upon patent expiration. However, because biologics are typically approved under the Public Health Service Act, these mechanisms were not available for most follow-on biotechnology products.

The Patient Protection and Affordable Care Act, P.L. 111-148, offers several provisions designed to encourage innovation in biotechnology while creating an accelerated approval process for a biosimilar drug. The legislation also permits limited market exclusivity for the first follow-on biologic to be designated "interchangeable" with the reference product. At the same time, the innovator biologic is afforded a 12-year data exclusivity period during which time the Food and Drug Administration will not approve a follow-on biopharmaceutical. In addition, the legislation establishes a specialized patent dispute resolution proceeding for biologic drugs.

Some experts maintain that the intellectual property provisions of P.L. 111-148 will foster both the development of new biotechnology products and the availability of lower-cost follow-on biologics. Other commentators argue that the need for additional safety and efficacy trials to test these products, and the fact that these drugs may be similar, but not identical, may add to the prices charged for the follow-on product. High manufacturing costs associated with biopharmaceuticals also may reduce significant cost savings from biosimilars.

Introduction

Congressional interest in the biotechnology industry parallels congressional concerns over the availability and cost of prescription drugs. The increasing significance of biologic pharmaceuticals to the health of the U.S. population has focused attention on this rapidly expanding market. Today, 20% of the available drugs are biologics[1] and some experts estimate that in the near future half of all newly approved pharmaceuticals will result from biotechnology.[2] However, biologics typically are more expensive than traditional, chemical-based drugs for multiple reasons including the high cost associated with research and development (R&D), clinical trials, and manufacturing. At the same time that these products have "contributed to increased life spans of today's population in much of the Western World," the innovations they have brought have contributed to the cost of health care.[3]

Biologics differ significantly from traditional pharmaceuticals in their complexity and method of manufacture. Typical pharmaceutical products consist of small molecules, on the order of dozens of atoms, that may be readily characterized and reproduced through well-understood chemical processes. In contrast, biologics are often made up of millions of atoms, feature a more complex structure than traditional pharmaceuticals, and are manufactured from living cells through biological processes.[4] The term "biologic" has been described as "poorly defined," and its precise parameters are themselves subject to debate.[5] The Public Health Service defines the term "biological product" to mean "a virus, therapeutic serum, toxin, antitoxin, vaccine, blood, blood component or derivative, allergenic product, or analogous product, or arsphenamine or derivative of arsphenamine (or any other trivalent organic arsenic compound), applicable to the prevention, treatment, or cure of a disease or condition of human beings."[6] Biologics are also sometimes termed "biotechnology drugs" or "biopharmaceuticals"[7] and typically are approved for marketing under the Public Health Service Act (PHS Act). Traditional, chemical drugs generally are approved under provisions of the Federal Food, Drug, and Cosmetic Act (FD&C Act).

P.L. 111-148, the Patient Protection and Affordable Care Act (H.R. 3590), provides mechanisms intended to foster innovation in biotechnology while facilitating the approval of follow-on biologic drugs, also known as "biosimilars." Congressional interest in the availability of lower- cost versions of innovative biotechnology drugs led to the creation of an accelerated approval process designed specifically for these follow-on products. The first biologic that is afforded "interchangeable" status with the reference drug is provided limited market exclusivity. Concurrently, the legislation furnishes the innovator biologic a 12-year data exclusivity period during which time the Food and Drug Administration (FDA) will not approve a follow-on product using safety and efficacy information provided by the original drug's sponsor. A specialized patent dispute resolution procedure is also established. This paper offers an overview of these provisions. For more detailed discussion on follow-on biologics see CRS Report RL33901, *Follow-On Biologics: Intellectual Property and Innovation Issues*, and CRS Report RL34045, *FDA Regulation of Follow-On Biologics*.

Expedited Marketing Approval

The marketing of a new drug is prohibited unless that product meets certain safety and efficacy standards. Sponsors of new drugs must submit documentation demonstrating that these standards have been met in order to obtain marketing approval. Typically, this is a complex, lengthy document that presents clinical data; chemistry, manufacturing and controls; nonclinical pharmacology and toxicology; safety update reports; and other salient information.[8] Brand-name drug companies commonly devote considerable resources, over many years, to complete the studies necessary to submit an application for a new pharmaceutical.

P.L. 111-148 creates an expedited approval process for follow-on biologics that are not able to utilize the accelerated approval mechanism established for traditional chemical drugs by the Hatch-Waxman Act.[9] Under the Hatch-Waxman Act, the manufacturer of a potential generic chemical drug may submit an Abbreviated New Drug Application (ANDA) to the FDA for approval if, generally speaking, the active ingredient, route of administration, the dosage form, and the strength of the new drug are the *same* as those of the approved drug. An ANDA allows a generic drug manufacturer to rely upon the safety and efficacy data of the original manufacturer. The availability of an ANDA typically permits a generic manufacturer to avoid the costs and delays associated with filing a full-fledged New Drug Application. ANDAs also may allow a generic manufacturer to place its FDA-approved bioequivalent drug on the market as soon as relevant patents expire.[10]

The complexity of biological products and the importance of the particular manufacturing process used to produce them may make the showing that a follow-on pharmaceutical is the "same" as a previously approved biologic—as required under the Hatch-Waxman Act—difficult, if not impossible.[11] Therefore, a somewhat different approach is taken in P.L. 111-148. This legislation creates a two-tiered expedited approval process based upon FDA determination of the type and breadth of studies necessary to determine whether a follow-on biologic is "biosimilar" or "interchangeable." Information provided to the FDA on the approval application must demonstrate that the follow-on biologic is *similar* to the reference product based on data from a clinical study or studies to demonstrate safety, purity, and potency for one or more appropriate conditions of use for which the reference product is licensed. In addition, the Secretary of Health and Human Services may determine that elements in the application, such as clinical studies, may be unnecessary.

The Secretary of Health and Human Services is to determine what data are to be provided in the application for licensing a follow-on biologic. A follow-on product may be designated as "biosimilar" if (1) analytical, animal, and clinical studies show that it is highly similar to the reference product, notwithstanding minor differences in clinically inactive components, (2) the two products have the same mechanism of action, (3) the condition of use in the proposed product has been previously approved for the reference product, (4) the route of administration, dosage form, and strength of the two products are the same, and (5) the manufacturing process provides for a safe product.

The Secretary may also designate a follow-on biologic as "interchangeable" if (1) it can be expected to produce the same clinical result as the reference product in any given patient and (2) the risk, in terms of safety or diminished efficacy or switching between the two products, is not greater than the use of the reference product without such alternation.

It is expected that the information required for a follow-on product approval will be less costly and take less time to generate than for the innovator drug.

Data Exclusivity

In an expedited approval process, the follow-on product is permitted to rely on the safety and efficacy data generated by the innovator drug (although only the FDA has actual access to the information). The Hatch-Waxman Act established a "data exclusivity" period during which the FDA will not consider competing applications for marketing approval of generic chemical drugs. Among the existing data exclusivities contained in this act is a five-year period available for drugs that qualify as new chemical entities. Should the original drug's sponsor submit new clinical studies in support, that sponsor may obtain a three-year period of data exclusivity that applies to the use of the product that was supported by the new clinical study.

At 12 years, P.L. 111-148 provides a more robust data exclusivity period for innovator biologics; no application for a biosimilar product may be approved for 12 years from the reference product's licensure date. No application for a follow-on biologic may be filed with the FDA for four years from the date the reference product was licensed. However, exclusivity will not be awarded for (1) supplements to the reference product application; (2) the identification of new indications, routes of administration, dosing, or delivery; or (3) modifications to the structure of the biological product that do not result in a change in safety, purity, or potency.

The 12-year exclusivity period is designed to address concerns that patents on biologics do not provide as potent a scope of proprietary rights as patents on small molecule drugs. Follow-on biologics cannot be the "same" as the reference product because they are made by different cell lines under different manufacturing processes. Thus, a firm utilizing an expedited approval process for a biosimilar drug based on the Hatch-Waxman Act would have access to the health, safety, and efficacy data of the innovator product but would be able to "design around" the reference drug without infringing the existing patents. Data exclusivity serves as an incentive to additional costly development and encourages venture capital investments necessary for continued R&D, testing, and commercialization. Proponents maintain that data exclusivity provides certainty in the process of developing a biologic which patents do not. This certainty generates competition in the marketplace where without significant data exclusivity for biotechnology products the venture capital sector is expected to invest elsewhere. In addition, commentators point out that data exclusivity does not prevent the introduction of a follow-on biologic product based on that company's own testing.

Market Exclusivity for First Interchangeable Product

Market exclusivity is awarded to the applicant that is the first to establish that its product is "interchangeable" with the reference product for any condition of use and is approved for use by the FDA. During this time, the FDA will not approve another follow-on product. The period of marketing exclusivity is the earlier of: (1) one year after the first commercial marketing of the first interchangeable biosimilar product to be approved as interchangeable with that reference product; (2) 18 months after either a final court judgment in patent infringement litigation under the Public Health Service (PHS) Act, as amended, or the dismissal of such litigation against the first applicant; (3) 42 months after the approval of the first interchangeable biosimilar biological product if patent litigation under the PHS Act, as amended, remains pending; or (4) 18 months after approval of the first interchangeable biosimilar product if the applicant has not been sued for patent infringement under the PHS Act, as amended.

Patent Dispute Resolution

In addition to creating expedited marketing approval pathways for generic drugs, the Hatch- Waxman Act incorporates numerous additional provisions pertaining to intellectual property.[12] Among these provisions are a statutory exemption from claims of patent infringement based on acts reasonably related to seeking FDA approval (commonly know as the "safe harbor"); patent term extension for a portion of the time spent seeking marketing approval; and special provisions for challenging the enforceability, validity, or infringement of approved drug patents.[13]

The applicability of the intellectual property provisions of the Hatch-Waxman Act to biologics presents complex issues. Some of the provisions of the Hatch-Waxman Act apply to biologics, whether they were approved under the FD&C Act or the PHS Act. In particular, the patent term extension and "safe harbor" provisions found in Title II of the Hatch-Waxman Act were enacted as general amendments to the Patent Act (Title 35 of the U.S. Code). The patent term extension statute, codified at 35 U.S.C. § 156, specifically accounts for the possibility of a "human biological product" approved under the Public Health Service Act. The "safe harbor" provision, found at 35 U.S.C. § 271 (e)(1), has been construed to apply to biologics as well.[14]

Congress framed the remaining intellectual property provisions of the Hatch-Waxman Act, including those establishing data exclusivities and specialized dispute resolution proceedings, as specific amendments to the FD&C Act. These terms would therefore apply to biologics only to the extent they were governed by the FD&C Act. If a particular biologic is approved under the auspices of the PHS Act, however, these provisions are not applicable. [15]

To address patent infringement issues for biologic drugs approved under the PHS Act, P.L. 111- 148 establishes a specialized patent dispute resolution proceeding that differs from the Hatch- Waxman framework. The legislation requires that within 20 days after the Secretary of Health and Human Services publishes a notice that its application has been accepted for review, the biosimilar or interchangeable product applicant will provide the reference product sponsor with details concerning the product and its production. Within 60 days of the date of receipt of that information, the reference product sponsor is then required to identify patents that it deems relevant to the biosimilar or interchangeable product. The reference product sponsor is required to update this list with any patents it subsequently acquires. The biosimilar or interchangeable product applicant is then allowed 60 days to provide the reference product sponsor with a list of patents it deems relevant to the reference product. The biosimilar or interchangeable product applicant also must state either that it will not market its product until the relevant patents have expired, or alternatively provide its views that the patents are invalid or would not be infringed. In the latter case, the reference product sponsor or interested third party is to provide the biosimilar or interchangeable product applicant with a response concerning the infringement and validity of those patents within 60 days.

The parties are required to engage in good faith negotiations to agree on which patents will be the subject of a patent infringement action. If those negotiations do not result in an agreement within 15 days, the parties will exchange lists of relevant patents to be litigated, with at least one patent identified by the reference product sponsor being subject to litigation. The reference product sponsor may then commence patent infringement litigation within 30 days. The biosimilar or interchangeable product applicant is to notify the reference product sponsor at least 180 days before it commences commercial marketing, and the reference product sponsor is allowed to seek a preliminary injunction at that time. Declaratory judgments are prohibited for either party unless the biosimilar or interchangeable product applicant fails to provide a product and patent application as required, at which time the reference product sponsor may seek a declaratory judgment.

Observations

Observers agree that the market for biologic pharmaceuticals is expanding by a number of measures including the quantity of approved products, the size of the market, and the importance of these drugs to the health of U.S. citizens. Patents on many biologics are expected to expire in the next few years; however, some commentators have expressed concerns that patent expirations may not be accompanied by the introduction of competing, lower-cost follow-on drugs in the marketplace. The intellectual property provisions of P.L. 111-148 are designed to address these concerns by establishing an accelerated approval process for follow-on products and creating a specialized patent dispute resolution mechanism. However, it remains to be determined if the cost savings will be as substantial as those generated by generic chemical drugs. The need for additional safety and efficacy trials to test these products, and the fact that these drugs may be similar, but not identical, may add to the prices charged for the follow-on product. Other factors may also affect the availability of biosimilars including high manufacturing costs and the rapid rate of innovation in the biotechnology sector. As the provisions of the new legislation become operable, experts will be better able to assess if the law has a significant effect on the availability of follow-on biologics while at the same time ensuring that innovation in the biotechnology arena continues.

Acknowledgments

This chapter was funded in part by a grant from the John D. and Catherine T. MacArthur

End Notes

[1] Ernst & Young, *Beyond Boarders, Global Biotechnology Report 2008*, 30, available at http://www.ey.com/ beyondborders.

[2] Herman Saftlas, "Industry Surveys, Healthcare: Pharmaceuticals," *Standard & Poor's*, June 4, 2009, 10, available at http://www.standardandpoors.com.

[3] Esther F. Schmid and Dennis A. Smith, "Pharmaceutical R&D in the Spotlight: Why is There Still Unmet Medical Need?" *Drug Discovery Today*, December 2007, 1001.

[4] Melissa R. Leuenberger-Fisher, "The Road to Follow On Biologics: Are We There Yet?," *Biotechnology Law Report,* August 2004, 389.

[5] David M. Dudzinski, "Reflections on Historical, Scientific, and Legal Issues Relevant to Designing Approval Pathways for Generic Versions of Recombinant Protein-Based Therapeutics and Monoclonal Antibodies," *Food and Drug Law Journal,* 2005, 143.

[6] 42 U.S.C. § 262(i) (2006)

[7] *Reflections on Historical, Scientific, and Legal Issues Relevant to Designing Approval Pathways for Generic Versions of Recombinant Protein-Based Therapeutics and Monoclonal Antibodies,* 143.

[8] Apotex Inc. v. FDA, 393 F.3d 210, 212 (D.C. Cir. 2004).

[9] 21 U.S.C. sec. 355 and following.

[10] Sarah E. Eurek, "Hatch-Waxman Reform and Accelerated Entry of Generic Drugs: Is Faster Necessarily Better?" *DUKE L. & TECH. REV.,* August 13, 2003.

[11] Debra Weintraub, "Next Generation of Biopharmaceuticals," *Journal of Managed Care Medicine*, 2006, available at http://www.namcp.com/Journals/JMCM/Articles/Next%20Generation%20of%20Biopharmaceuticals.pdf.

[12] P.L. 84-417, 98 Stat. 1585 (1984).

[13] A more detailed discussion of the Hatch-Waxman Act is found in CRS Report R41 114, *The Hatch-Waxman Act: A Quarter Century Later*, by Wendy H. Schacht and John R. Thomas andRL3 0756, Patent Law and Its Application to the Pharmaceutical Industry: An Examination of the Drug Price Competition and Patent Term Restoration Act of 1984 ("The Hatch-Waxman Act"), by Wendy H. Schacht and John R. Thomas.

[14] *Patent Fairness Act of 1999: Hearing on H.R. 1598,* House Comm. on the Judiciary, 106th Cong. 94, 1999, Statement of Hon. Henry A. Waxman, Member, House Comm. on Government Reform, available at http://commdocs.house.gov/ committees/judiciary/hju62499.000/hju62499_0f.htm.

[15] Terry Mahn and Margo Furman, *The Role of Patent and Non-Patent Exclusivity Under the Hatch-Waxman Act,* 2005, 19-20, available at http://www.fr.com/practice/pdf/05-16-05%20_PatentandNon-PatentExclusivity.pdf.

In: Follow-On Biologics: Background and Issues
Editor: Daniel J. Gutiérrez

ISBN: 978-1-61122-076-6
© 2011 Nova Science Publishers, Inc.

Chapter 3

Follow-On Biologics: Intellectual Property and Innovation Issues

Wendy H. Schacht and John R. Thomas

Summary

Biologics, which are sometimes termed biopharmaceutials or biotechnology drugs, have begun to play an increasingly important role in U.S. health care. Not only are sales of biologics growing rapidly, some experts estimate that in coming years half of all newly approved drugs will result from biotechnology.

A number of patents pertaining to certain biological products will expire in the near future. Some congressional concern has been voiced over the possibility that these patent expirations may not be accompanied by the introduction of competing, lower-cost biologics in the marketplace. With respect to traditional pharmaceuticals, the Drug Price Competition and Patent Term Restoration Act of 1984, a statute commonly known as the "Hatch-Waxman Act," is widely believed to have encouraged the availability of generic substitutes for brand-name pharmaceuticals upon patent expiration. The Act in part permitted the Food and Drug Administration to expedite its marketing approval proceedings with respect to generic drugs.

Some observers believe that the Hatch-Waxman Act's accelerated marketing approval provisions do not comfortably apply to biologics, however. Biologics differ significantly from traditional small-molecule pharmaceuticals in their size, structural complexity, and method of manufacture. Competitors who wish to develop follow-on biologics may face difficult, and even insurmountable difficulties in demonstrating that their product is equivalent to a particular brand- name biologic. Other commentators assert that different kinds of biologics vary considerably in their size and structure, and believe that existing Hatch-Waxman mechanisms are appropriately applied to many biologics.

The patent system also plays a role in regulating competition in the biologics market. Patent protection is available for biologics in many circumstances, although the scope of protection may be limited by legal principles that restrict the availability of proprietary rights in naturally occurring substances.

In the 111th Congress, H.R. 3962 (passed by the House on November, 14, 2009) and H.R. 3590 (passed by the Senate on December 24, 2009), as well as S. 1679, H.R. 3200, H.R. 1427, S. 726, and H.R. 1548, while varying in details, would create an expedited marketing approval pathway for follow-on biologics within the Public Health Service Act. These bills also would establish specialized patent dispute resolution proceedings with respect to these products. Data and market exclusivity periods for innovator drugs and follow-on products are also created.

Introduction

Longstanding congressional interest in the availability and cost of pharmaceuticals has focused attention upon the increasingly significant class of drugs known as "biologics."[1] Observers agree that the biologics market is rapidly expanding by any number of measures, including the quantity of approved products, the size of the market, and the importance of these drugs to the health of U.S. citizens. The Food and Drug Administration (FDA) issued marketing approval on 36 biotechnology drugs in 2002; it also approved 37 in the following year, 40 in 2004, 38 in 2005, and 36 in 2006.[2] While the number of approvals declined in 2007, and remained below previous years in 2008,[3] many more new biologics reportedly are in the pipeline and/or in the approval process.[4] Today, 20% of the drugs on the market are biologics.[5] Federico Polliano, head of business development at BioGeneriX, a biotechnology company, projects that 50% of approved pharmaceuticals in 2010 will be the result of biotechnology.[6]

Dramatic growth in the number of approved drugs has been accompanied by a similar expansion in sales. IMS Health, a consulting firm, found that in 2005, the size of the U.S. biologics market was on the order of \$52 billion. According to its analysis, the biologics market grew at a rate of 17%, greater than any other portion of the pharmaceutical market.[7] Some experts further project that the global biologics market will expand to \$67 billion by 2010.[8] Awareness of the increasing importance of biopharmaceuticals has been accompanied by an appreciation that patents covering many of these products will soon expire. Andrew Forman of WR Hambrecht concludes, for example, that \$20 billion in biotech drugs worldwide will be off patent by 2010.[9]

Estimates vary on the number of biologics that will lose patent protection in the next several years and the amount of sales these products represent. One study notes that "between 2009 and 2019 21 blockbuster biopharmaceuticals with a total market value of over 50 billion \$US will lose patent protection."[10] Analysis by FierceBiotech estimates that biologics worth \$25 billion in sales will lose patent protection by 2016.[11] A report by Price Waterhouse Coopers notes that between 2002 and 2013, the expiration of patents on blockbuster biopharmaceutical products will lead to an average loss of \$16.4 billion in sales.[12] Among the biotechnology drugs that are expected to go off patent between 2008 and 2010 are Genotropin, NovoSeven, BeneFIX, Infergen, and Humira (\$455,911,000 in 2008 sales by retail dollars[13]).[14] Patents on Avonex (\$273,303,000 in 2008 sales by retail dollars[15]) and

Enbrel ($ 763,927,000 in 2008 sales by retail dollars[16]) are also expected to expire in the next several years.[17]

Some commentators have expressed concerns that patent expirations may not be accompanied by the introduction of competing, lower-cost biologics in the marketplace.[18] In the traditional pharmaceutical market, generic substitutes commonly become available to consumers as patents on brand-name drugs expire due to the provisions of the Drug Price Competition and Patent Term Restoration Act of 1984, a statute commonly known as the "Hatch-Waxman Act."[19] This legislation introduced several significant changes to both the patent law and the food and drug law established by the Federal Food, Drug, and Cosmetics Act. Among them were expedited marketing approval pathways that eliminated, in whole or in part, the need for firms to conduct expensive and time-consuming clinical trials when they bring generic equivalents of brand-name drugs to market.[20] The Hatch-Waxman amendments were designed to facilitate the rapid introduction of lower-cost generic drugs, while at same time promoting innovation in the pharmaceutical industry.[21]

Technical factors may limit the effectiveness of the Hatch-Waxman amendments to biologics, however. Biologics differ significantly from traditional pharmaceuticals in their complexity and method of manufacture. Typical pharmaceutical products consist of small molecules, on the order of dozens of atoms, that may be readily characterized and reproduced through well-understood chemical processes. In contrast, biologics are often made up of millions of atoms, feature a more complex structure than traditional pharmaceuticals, and are manufactured from living cells through biological processes.[22] As a result, the technical challenges that a competitor faces in developing a product that may be viewed as equivalent to a particular brand-name biologic product may be considerable, and in some cases perhaps even insurmountable.[23] For this reason, many experts do not describe competing biologic products as "generics," as is the case for a small-molecule pharmaceuticals; the terms "follow-on biologic" or "biosimilar" are commonly used instead. [24]

Some commentators assert that these technical challenges may also mean that the expedited approval pathways available under the Hatch-Waxman Act do not comfortably apply to biologics most of which are approved under provisions of the Public Health Services Act (PHS Act). Because the complexity of biologics is an order of magnitude greater than that associated with pharmaceuticals, they say, an expedited marketing approval protocol would not ensure patient safety to the degree possible with respect to traditional drugs.[25] Others observe that different kinds of biologics vary considerably in their size and structure, and believe that existing Hatch- Waxman mechanisms provide appropriate regulatory oversight for less complex biologics and also should be extended to those biologics approved under the PHS Act. These observers further explain that as scientific knowledge progresses, understanding of biologics will increase, thereby allowing expanded use of current procedures.[26]

FDA marketing approval is not the only gatekeeper to competition in the biologics market. The patent system also has a role to play. Although patent protection may be available for biologics in many circumstances, these patents may be limited by legal principles that restrict proprietary rights in naturally occurring substances.[27] Further, although key patents on many biologics are set to expire, other products may potentially remain protected by patents for many years to come. In recognition that the public possesses an interest in prompt challenges to drug patents that are believed to have been improvidently granted, the Hatch-Waxman Act introduced incentives for firms to bring such challenges along with special procedures for resolving them in the courts.[28]

This chapter reviews doctrinal and policy issues pertaining to follow-on biologics. The report first introduces the application of federal food and drug legislation to follow-on biologics. It next turns to the patent implications of marketing follow-on biologics. Following this review of substantive law, the remainder of the report introduces innovation policy issues pertaining to follow-on biologics.

Expedited Marketing Approval Issues

Efforts to speed approval of follow-on biologics have been based upon the Hatch-Waxman Act. [29] Enacted in 1984, that statute amended the Federal Food, Drug, and Cosmetic Act (FDC Act) in part by introducing two regulatory pathways allowing for the expedited marketing approval of generic pharmaceuticals. Many policy and industry experts agree that the Hatch-Waxman Act has significantly effected the availability of generic substitutes for brand name, small molecule drugs.[30] Generics are often rapidly available after patent expiration, commonly at lower prices than the brand-name original. Concurrently, given the increasing investment in research and development (R&D) and the gains in research intensity of the pharmaceutical industry, it appears that the 1984 Act has not deterred the search for, and the development of, new drugs.[31]

At issue is whether the existing Hatch-Waxman Act protocols, or a similar construct created through new legislation, could be effectively implemented for follow-on biologics. The difficulty characterizing biologics, along with the importance of manufacturing techniques to the final product, have created debate around the applicability of expedited marketing approval mechanisms for these products. This chapter next describes issues pertaining to the accelerated marketing approval of follow-on biologics.

Expedited Marketing Approval for Pharmaceuticals

The FDC Act has, since 1962, prohibited the marketing of a "new drug" unless that drug meets certain safety and efficacy standards. Sponsors of new drugs must submit, among other documents, a New Drug Application (NDA) demonstrating that these standards have been met in order to obtain marketing approval. A typical NDA is a complex, lengthy document that presents clinical data; chemistry, manufacturing and controls; nonclinical pharmacology and toxicology; safety update reports; and other salient information.[32] Brand-name drug companies commonly devote considerable resources, over many years, to complete the studies necessary to submit a NDA.

Prior to the passage of the Hatch-Waxman Act, the federal food and drug law contained no separate provisions addressing generic versions of drugs that had previously been approved. [33] The result was that a would-be generic drug manufacturer most often had to file its own NDA in order to market its drug.[34] Some generic manufacturers were forced to prove independently that the drugs were safe and effective, even though their products were chemically identical to those of previously approved drugs. Some commentators believed that the approval of a generic drug was a needlessly costly, duplicative, and time-consuming process prior to the Hatch-Waxman Act.[35] These observers noted that although patents on important drugs had expired, manufacturers were not moving to introduce generic equivalents for these products due to the level of resource expenditure required to obtain FDA marketing approval.[36]

In response to this concern, the Hatch-Waxman Act created two new types of applications for marketing approval of a generic pharmaceutical. One is termed an "Abbreviated New Drug Application" (ANDA).[37] An ANDA may be filed, generally speaking, if the active ingredient, route of administration, the dosage form, and the strength of the new drug are the same as those of the approved drug. An ANDA allows a generic drug manufacturer to rely upon the safety and efficacy data of the original manufacturer. The availability of an ANDA typically permits a generic manufacturer to avoid the costs and delays associated with filing a full-fledged NDA. ANDAs also allow a generic manufacturer, in many cases, to place its FDA-approved bioequivalent drug on the market as soon as any relevant patents expire.[38]

The Hatch-Waxman Act also authorized a so-called " § 505(b)(2) application." A § 505(b)(2) application is one for which one or more of the investigations relied upon by the generic applicant for approval "were not conducted by or for the applicant and for which the applicant has not obtained a right of reference or use from the person by or for whom the investigations were conducted.... "[39] A § 505(b)(2) application differs from an ANDA in that it includes full reports of investigations concerning the safety and effectiveness of the proposed product. However, a § 505(b)(2) application is distinct from an NDA in that the § 505(b)(2) application relies upon data that the applicant did not develop.[40]

According to the FDA, a § 505(b)(2) applicant can rely upon two sources of studies "not conducted by or for the applicant and for which the applicant has not obtained a right of reference."[41] The first source consists of safety and efficacy analysis based upon data that the applicant did not originate itself and does not enjoy an express permission to access. This category of information typically consists of published scientific literature. As a result, § 505(b)(2) applications are sometimes referred to as "paper NDAs."

With respect to the second source of appropriate information, the FDA has declared that a § 505(b)(2) applicant may rely upon that agency's own finding of safety and effectiveness for an approved drug.[42] The FDA's conclusion allows applicants that wish to market a modification of an approved drug to file a § 505(b)(2) application rather than a full NDA. For example, suppose that an approved drug employed an active ingredient with a particular salt formulation. A generic firm seeks to market a generic version of the approved drug with the same active moiety, but using a different salt formulation.[43] Under these circumstances, the generic firm may be unable to file an ANDA because its proposed active ingredient is not identical to that of the approved product.[44] Due to the FDA's view, however, the generic firm could file a § 505(b)(2) application that relied upon the FDA's previous approval of the innovator drug, along with studies supporting the change in the salt formulation from the innovator drug.[45] As may be appreciated, the availability of a § 505(b)(2) application likely leads to a substantial reduction in the costs of the generic firm in this case in comparison with the resources required to file a full NDA.

The FDA interpretation of § 505(b)(2) regarding this second source of information has been criticized. Some observers believe that under a plain reading of the Hatch-Waxman Act, a generic applicant may rely upon an innovator's proprietary data only when filing an ANDA application, not a § 505(b)(2) application. Under this view, the FDA position inappropriately expands the circumstances to which a § 505(b)(2) application applies.[46] Despite this critique, the FDA has taken the position that its "approach is intended to encourage innovation in drug development without requiring duplicative studies to demonstrate what is already known about a drug while protecting the patent and exclusivity rights for the approved drug."[47]

Application to Biologics

The Hatch-Waxman Act, including its expedited approval pathways, may potentially apply to certain biologics. This result follows because the Hatch-Waxman Act in part amended the FDC Act, which in turn applies to "drugs."[48] The Hatch-Waxman Act did not amend a distinct statute, the Public Health Service Act (PHS Act),[49] which applies specifically to biological products and contains no provisions allowing expedited marketing approval for follow-on biologics. Because the definition of "drugs" under the FDC Act is broad, however, the FDA states that "[b]iological products subject to the PHS Act also meet the definition of drugs under the Federal Food, Drug, and Cosmetic Act."[50] Under this interpretation, both expedited marketing approval pathways—the ANDA and the § 505(b)(2) application—could potentially be available for pharmaceuticals and biologics alike.[51]

Many observers believe that the possibility of employing an ANDA for a biologic is currently more theoretical than real, however, due to differences between this class of drugs and conventional pharmaceuticals. Two of the more significant differences are the increased complexity of biologics vis-à-vis pharmaceuticals, as well as the importance of the particular manufacturing process employed to produce a biologic. Food and drug lawyers have stated:

> Biologics differ significantly from traditional drugs in their size, complexity, structure, and method of manufacture. Drug products consist of small molecules, on the order of dozens of atoms synthesized from defined components according to a prescribed production method in an environment of manufacturing processes and controls. Biological products are much larger than drugs, made up of millions of atoms, and are manufactured from living cells through an elaborate process initiated by specifically programming a cell line to produce a certain protein in a highly controlled, sterile manufacturing environment.[52]
>
> Characteristic properties of biologics include a very high molecular weight and high structural complexity, a heterogeneous molecular make-up, varying levels of hard-to-remove biological impurities (bacteria, viruses and the like) and a high degree of sensitivity to environmental conditions.[53]

These traits of biologics make the manufacturing process critical to the final product. It has been argued that "[t]he production process is 90 percent of the intellectual property related to the product."[54] In contrast to traditional drugs, which are manufactured by chemical synthesis,[55] the production of biologics is more complex:

> [M]anufacturing biologics requires the nourishment and support of living host cells transfected with genetically engineered DNA to code for the desired biological protein that is expressed by the host cell. A master cell bank must be established, and host cells are cultured and fermented in large-scale bioreactors to produce commercial quantities of the desired protein.[56]

Not only are the characteristics of a biologic "clearly dependent on the process used to manufacture the product," such information is ordinarily protected via trade secrets.[57] As a result, some observers have opined that a "manufacturer would have great difficulty producing a follow- on protein that is identical to the innovator product."[58]

The complexity of biological products and the importance of the particular manufacturing process used to produce them may make the showing that a follow-on product is the "same" as a previously approved biologic difficult, if not impossible.[59] Former Acting FDA Commissioner Lester Crawford has explained that "because protein drug products are large, complex molecules, derived from biological sources, generally it has not been possible to assess relative sameness with a high degree of confidence."[60] Some observers have gone further, opining that true "generic" biologics cannot exist because they cannot be judged to be the "same"as the brand- name product, a requirement that the Hatch-Waxman Act imposes with respect to an ANDA.[61]

Other commentators believe that scientific capabilities currently allow manufacturers to produce follow-on biologics that are safe and effective.[62] For example, the Generic Pharmaceutical Association (GPhA)

> strongly believes FDA's current statutory structure would permit approvals and marketing of an array of generic biopharmaceuticals (also referred to as 'biologics') with relatively low to moderate complexity, and to expand [sic] that system in the coming years to permit the approval of more complex products as the science evolves.[63]

Israel Makov, president and CEO of Teva, argues that biologic products can be adequately characterized and manufacturing processes do not necessarily affect the final drug.[64] Similarly, Patrick Vink, Global Head of Biopharmaceuticals at Sandoz, maintains both that the science is in place to assess follow-on biologics appropriately, and that unnecessary clinical trials should not be mandatory. Instead, he advocates that bioequivalence testing should be required on a case-by-case basis.[65]

In this regard, some commentators have suggested that potential manufacturers of follow-on biologics employ § 505(b)(2) applications in appropriate cases. The statutory requirements for § 505(b)(2) applications do not require that the follow-on product be the "same" as the approved product. Rather, the applicant must provide clinical data demonstrating the follow-on product to be safe and effective, but may rely upon data generated by the brand-name firm itself, or by third parties.[66] The applicant must substantiate the "relevance and applicability" of previous findings to the follow-on product, however, and may need to supply clinical data to describe any relevant distinctions between the brand-name biologic and the follow-on product.[67]

It should be appreciated that some observers believe that § 505(b)(2) applications are inappropriate for biologics, however. For example, the Biotechnology Industry Organization (BIO) has asserted that:

> Approval of follow-on biotechnology products must be based on the same rigorous standards applied by the FDA for the approval of pioneer biotechnology products.... Currently, the science does not exist to provide an alternative to a full complement of data, including clinical evidence, to demonstrate safety and effectiveness for follow-on biotechnology products.... Therefore, in the current state of scientific knowledge and

technique, a clinical trial remains a fundamental principle for evaluating the safety and effectiveness of a follow- on biotechnology product.[68]

Legislative Initiatives

In the 111th Congress various bills have been introduced that would create an expedited approval process for follow-on biologics.[69] H.R. 3962, the Affordable Health Care for America Act, as passed by the House on November 14, 2009, and H.R. 3590, the Patient Protection and Affordable Care Act, as passed by the Senate on December 24, 2009, as well as S. 1679, the Affordable Health Choices Act; H.R. 3200, America's Affordable Health Choices Act of 2009; and H.R. 1548, the Pathway for Biosimilars Act, mandate that a follow-on biologic may be designated as either a "biosimilar" or an "interchangeable" product. In general, a follow-on product is biosimilar if (1) analytical, animal, and clinical studies show that it is highly similar to the reference product, notwithstanding minor differences in clinically inactive components, (2) the two products have the same mechanism of action, (3) the condition of use in the proposed product has been previously approved for the reference product, (4) the route of administration, dosage form, and strength of the two products are the same, and (5) the manufacturing process provides for a safe product. A follow-on biologic is interchangeable if (1) it can be expected to product the same clinical result as the reference product in any given patient and (2) the risk, in terms of safety or diminished efficacy or switching between the two products, is not greater than the use of the reference product without such alternation.[70]

H.R. 1427 and S. 726, both titled the Promoting Innovation and Access to Life-Saving Medicine Act, establish an expedited process for a biosimilar when "no clinically meaningful differences between the biological product and the reference product would be expected in terms of the safety, purity, and potency if treatment were to be initiated with the biological product instead of the reference product." A designation of "interchangeable" can be made if the follow-on product is determined to be biosimilar and can be switched one or more times with the original biologic without "an expected increase in the risk of adverse effects.... " The Secretary of Health and Human Services can issue a "biosimiliar biological product license" if the information provided meets the criteria established in the legislation, and determined to be necessary by the Secretary including (1) similar molecular structural features; (2) clinical and non-clinical studies indicating saftey, purity, and potency; (3) previous approval of the reference product; (4) same route of administration, dosage form, and strength as the original biologic; and (5) safe manufacture facilities, among other things.[71]

Intellectual Property Issues

Patent Protection for Biologics

As with pharmaceuticals, biologics may be subject to patent protection provided certain conditions are met.[72] An award of marketing approval by the FDA and the grant of a patent by the U.S. Patent and Trademark Office (USPTO) are distinct events that depend upon different criteria. FDA procedures determine whether the drug is sufficiently safe and effective to be marketed. In contrast, the USPTO grants patents on inventions that fulfill requirements established by the Patent Act of 1952, including utility, novelty, and nonobviousness.[73] Because patent proprietors may be able to block competitors during the term of the patent,[74] these intellectual property rights also play a role in the marketplace availability of follow-on biologics.

Although a complete review of the patent system exceeds the scope of this chapter, its basic contours may be concisely stated. The Patent Act allows inventors to obtain patents on processes, machines, manufactures, and compositions of matter that are useful, novel and nonobvious.[75] An invention is judged as useful if it is minimally operable towards some practical purpose.[76] To be considered novel within the patent law, an invention must differ from existing references that disclose the state of the art, such as publications and other patents.[77] The nonobviousness requirement is met if the invention is beyond the ordinary abilities of a skilled artisan knowledgeable in the appropriate field. [78]

In order to receive a patent, an inventor must file a patent application with the USPTO.[79] Patent applications must include a specification that so completely describes the invention that skilled artisans are enabled to practice it without undue experimentation. [80] The patent application must also contain distinct, definite claims that set out the proprietary interest asserted by the inventor.[81]

Trained personnel at the USPTO, known as examiners, review all applications to ensure that the invention described and claimed in the application fulfills the pertinent requirements of the patent law. If the USPTO believes that the application fulfills the statutory requirements, it will allow the application to issue as a granted patent.[82] Each patent ordinarily enjoys a term of twenty years commencing from the date the patent application was filed.[83] If the patent proprietor was unable to market its product for a period of the patent term due to lack of approval by the FDA, the term may be extended by a portion of the regulatory review period in some circumstances.[84]

Granted patents give the patentee the right to exclude others from making, using, selling, offering to sell, or importing into the United States the patented invention.[85] Parties who engage in those acts without the permission of the patent proprietor during the term of the patent can be held liable for infringement. The patentee may file a civil suit in federal court in order to enjoin infringers and obtain monetary remedies.[86] Although issued patents enjoy a presumption of validity, accused infringers may assert that the patent is invalid or unenforceable on a number of grounds.[87]

A few patent law principles have particular impact upon biologics. First, the courts have generally concluded that novel and nonobvious products and processes of biotechnology may be patented, notwithstanding the fact that the invention derives from the field of biochemistry or is itself a "living invention." In *Diamond v. Chakrabarty*,[88] for example, the Supreme Court held a genetically modified bacterium was patentable. The Court explained that the inventor's "claim is not to a hitherto unknown natural phenomenon, but to a nonnaturally occurring manufacture or composition of matter—a product of human ingenuity 'having a distinctive name, character [and] use.'"[89]

Biotechnology firms may at times confront the longstanding patent law principle that a naturally occurring substance may not be patented as such.[90] For example, a scientist could not obtain a patent on a previously unknown plant that she discovered growing in the wild. A patent may be obtained once significant artificial changes are made to that natural substance, however. For example, a biological substance that is discovered in nature and isolated from its source may be subject to patent protection. Amgen's Epogen®, a genetically engineered form of erythropoietin that combats anemia by encouraging the production of red blood cells, provides one example of a patented biologic.[91]

In addition, a patent may be available for a new process used to manufacture a known biologic.[92] Suppose, for example, that a naturally occurring biological agent and its activity are already known to the state of the art. The contribution of the biotechnology firm is to develop a manufacturing process that allows for widespread, commercial use of that agent. In this scenario, the biotechnology firm may obtain a patent on the manufacturing process, but not upon the biological agent itself. It should be appreciated that the value of such a process patent will depend upon whether competitors will be able to employ a distinct manufacturing process in order to create a comparable final product. Because of the potential ability of competitors to design around process patents, some observers believe that process patents may be of less significance in the marketplace than patents directed towards the final product itself.[93]

The Hatch-Waxman Act, Intellectual Property, and Biologics

In addition to creating expedited marketing approval pathways for generic drugs, the Hatch- Waxman Act incorporated numerous additional provisions pertaining to intellectual property.[94] Among these provisions are a statutory exemption from claims of patent infringement based on acts reasonably related to seeking FDA approval (commonly know as the "safe harbor"); patent term extension for a portion of the time spent seeking marketing approval; special provisions for challenging the enforceability, validity, or infringement of approved drug patents; data exclusivities for brand-name firms; and a reward for challenging patent enforceability, validity, or infringement consisting of 180 days of generic exclusivity to the first generic applicant to file a patent challenge against any approved drug.[95]

The applicability of the intellectual property provisions of the Hatch-Waxman Act to biologics presents complex issues. Some of the provisions of the Hatch-Waxman Act plainly apply to biologics, whether they were approved under the FDC Act or the PHS Act. In particular, the patent term extension and "safe harbor" provisions found in Title II of the Hatch-Waxman Act were enacted as general amendments to the Patent Act (Title 35 of the U.S. Code). The patent term extension statute, codified at 35 U.S.C. § 156, specifically accounts for the possibility of a "human biological product" approved under the Public Health Service Act. The "safe harbor" provision, found at 35 U.S.C. § 271(e)(1), has been construed to apply to biologics as well.[96]

Congress framed the remaining intellectual property provisions of the Hatch-Waxman Act, including those establishing data exclusivities and specialized dispute resolution proceedings, as specific amendments to the FDC Act. [97] These provisions would therefore apply to biologics only to the extent they were governed by the FDC Act. To the extent that a particular biologic is approved under the auspices of the PHS Act, however, these provisions would be inapplicable.[98]

Should the FDC Act apply to a particular biologic, one of its more notable intellectual property provisions relates to so-called data exclusivities (sometimes called "marketing exclusivities"). The term "data exclusivity" refers to a period of time during which the FDA affords an approved drug protection from competing applications for marketing approval. Among the existing data exclusivities under the Hatch-Waxman Act is a five-year period available for drugs that qualify as new chemical entities. Should the drug's sponsor submit new clinical studies in support, that sponsor may obtain a three-year period of data exclusivity that applies to the use of the product that was supported by the new clinical study.

Data Exclusivities

If a new, expedited FDA approval process is legislated for follow-on biologics under the Public Health Service Act, the length of any data exclusivity period appears to be the subject of debate. Some commentators, including the Generic Pharmaceutical Association (GPhA), argue the Hatch- Waxman Act provides a good model for the treatment of biosimilars in that the competition engendered by traditional generic drugs under the Act has "provided the U.S. with the most robust pharmaceutical industry in the world."[99] Some experts tend to dismiss the idea of added data exclusivity provisions as a necessary incentive for the development of new biologics and note that there is no data to support a need for any incentives beyond those associated with the Hatch-Waxman Act. According to GPhA,

> there is no reasonable justification for granting unprecedented and excessive market exclusivity periods to biopharmaceuticals. Any market exclusivity terms that are longer than five years diminish the importance of and reliance upon our patent system. Providing long government-granted absolute, market exclusivity terms would permit the developer of biopharmaceuticals to "double dip" by taking advantage of the absolute market exclusivity provision as well as garnering the fruits of the patent system.

Patents are viewed by many as sufficient to protect the intellectual property of innovator biologic firms and patent protection is the same for traditional pharmaceuticals as it is for biologics. Opponents of additional data exclusivity periods maintain that while patents can be challenged in court, an award of data exclusivity is effectively incontestable as stated in the response to an April 3, 2008 questionnaire from the House Committee on Energy and Commerce by Barr Pharmaceuticals, Inc. Some commentators contend that significantly extending data exclusivity for reference biologics would result in a guaranteed monopoly longer than that provided under the Hatch-Waxman Act. In addition, Barr stated that there are a "multitude of financial incentives" for new biologic product development including: patent term restoration; U.S. Patent and Trademark Office restoration for delays in patent approval; orphan drug exclusivity; orphan drug tax credits; general business research and experimentation tax credits; Puerto Rico activity tax credit; foreign tax credit; and Uruguay Round Agreement Act patent term restoration.

Other experts maintain that the Hatch-Waxman model is not appropriate for follow-on biologics. These commentators assert that patents on biologics do not provide as robust a scope of proprietary rights as patents on small molecule drugs. Follow-on biologics cannot be the "same" as the reference product because they are made by different cell lines under different manufacturing processes. Therefore, it is argued, the patent-related provisions of the Hatch- Waxman Act do not protect the brand name companies' investments as they would with traditional chemical drugs. As stated by biopharmaceutical company Millennium Pharmaceuticals, Inc.,

> due to the differences between small molecule drugs and biologics, the scope of protection afforded by a patent is narrower for a biologic than a traditional drug. With respect to small molecule drugs [e.g. chemical drugs], the key patent protection is on the active pharmaceutical ingredient, which, if patented, cannot be copied by a competitor during the term of protection. A patent on an innovator biologic, however, may not be infringed by a "highly similar" biologic. Thus, although patents for biologics provide critical incentives, particularly during the period of initial discovery and development, the protection that they offer should not be compared to that of the patents on traditional small molecule drugs.

Thus, a firm utilizing an expedited approval process for a biosimilar drug based on the Hatch- Waxman Act would have access to the health, safety, and efficacy data of the innovator product but would be able to "design around" the reference drug without infringing the existing patents.

Many argue that additional data exclusivities are necessary to encourage innovative new biologics. Data exclusivity serves as an incentive to additional costly development and encourages venture capital investments necessary for continued R&D, testing, and commercialization. In addition, proponents maintain that data exclusivity provides certainty in the process of developing a biologic which patents do not. This certainty generates competition in the marketplace where without significant data exclusivity for biotechnology products the venture capital sector is expected to invest elsewhere. In addition, commentators point out that data exclusivity does not prevent the introduction of a follow-on biologic product based on the generic company's own testing. According to Amgen, Inc.,

> A patent would prevent another manufacturer from marketing the same drug (or pay damages for infringement), even with a full application supported by its own data. Data exclusivity simply prevents reliance on the innovator's date for FDA approval, but does not prohibit a manufacturer from gaining approval of a product based on their own data.

Legislative Activity

Legislation introduced in the 111th Congress would amend the Public Health Service Act in order to provide specialized patent dispute resolution proceedings for biologics. The proposed regimes would act differently than the existing Hatch-Waxman framework. To date, H.R. 3962, the Affordable Health Care for America Act, as passed by the House on November 14, 2009, and H.R. 3590, the Patient Protection and Affordable Care Act, as passed by the Senate on December 24, 2009, include such provisions. (See **Table 1** below for a comparison of the relevant sections contained in H.R. 3962 and H.R. 3590.)

H.R. 3962 (as passed by the House), H.R. 3200, and H.R. 1548 mandate that within 30 days of the filing of a biosimilar application, the Secretary publish a notice of the application. The follow- on applicant also must provide the reference product owner with information concerning the biosimilar product and its manufacture. Within 60 days of the receipt of this information, the reference product sponsor must identify patents that it deems relevant to the biosimilar product. The reference product sponsor is required to update this list with any patents it subsequently acquires. In addition, "interested third parties" may inform the biosimilar applicant of their relevant patents at any time, and then be entitled to obtain information concerning the follow-on product and its production. The interested third party is then provided 90 days to identify further relevant patents; as with the reference product sponsor, the interested third party must update this list with subsequently acquired patents. Reference product sponsors and interested third parties must additionally state why they believe the biosimilar product would infringe their patents. The biosimilar applicant is then allowed 45 days to state either that it will not market its product until the relevant patents have expired, or alternatively provide its views that the patents are invalid or would not be infringed. In the latter case, the reference product owner of the interested third party may commence a patent infringement lawsuit against the biosimilar applicant within 60 days. Under H.R. 1548, follow-on applicants would be unable to seek declaratory judgment with respect to patents that are subject to notice and explanation by their owners until the later of 120 days after the explanation is provided, or 3 years prior to the expiration of marketing exclusivity.[100]

H.R. 3962 also would amend the Patent Act to stipulate that the filing of a statement by a biosimilar applicant regarding patents identified by the reference product sponsor and other interested parties may be considered an act of patent infringement. This legislation would require reference product sponsors, and allow other interested parties, to identify patents that relate to the proposed biosimilar product. The biosimilar applicant would then be afforded the opportunity to state its position regarding those patents. If the biosimilar applicant responds by asserting that one or more of these patents are invalid, unenforceable, or would not be infringed by the proposed biosimilar product, the biosimilar applicant would be deemed to have committed an act of patent infringement that would be immediately actionable in the courts.

H.R. 3590 (as passed by the Senate) and S. 1679 require that within 20 days after the Secretary publishes a notice that its application has been accepted for review, the biosimilar or interchangeable product applicant will provide the reference product sponsor with details concerning the product and its production. Within 60 days of the date of receipt of that information, the reference product sponsor is then required to identify patents that it deems relevant to the biosimilar or interchangeable product. The reference product sponsor is required to update this list with any patents it subsequently acquires. The biosimilar or interchangeable product applicant is then allowed 60 days to provide the reference product sponsor with a list of patents it deems relevant to the reference product. The biosimilar or interchangeable product applicant must also state either that it will not market its product until the relevant patents have expired, or alternatively provide its views that the patents are invalid or would not be infringed. In the latter case, the reference product sponsor or interested third party must provide the biosimilar or interchangeable product applicant with a response concerning the infringement and validity of those patents within 60 days. The parties are then required to engage in good faith negotiations to agree on which patents will be the subject of a patent infringement action. If those negotiations do not result in an agreement within 15 days, the parties will exchange lists of relevant patents to be litigated, with at least one patent identified by the reference product sponsor being subject to litigation. The reference product sponsor may then commence patent infringement litigation within 30 days. The biosimilar or interchangeable product applicant must notify the reference product sponsor at least 180 days before it commences commercial marketing, and the reference product sponsor is allowed to seek a preliminary injunction at that time. Declaratory judgments are prohibited for either party unless the biosimilar or interchangeable product applicant fails to provide a product and patent application as required, at which time the reference product sponsor may seek a declaratory judgment.

Follow-on or prospective follow-on applicants are permitted to request identification of all relevant patents from the holder of the reference product under H.R. 1427 and S. 726. The brand- name firm would then be required to respond to such a request within 60 days. The reference product holder further is mandated to update this list for a period of two years.[101] The follow-on applicant is able to challenge any identified patent by providing the basis for the challenge to the patent proprietor and holder of the reference product. The patent proprietor is required to bring a patent infringement suit within 45 days of notice of a challenge or lose the right to certain remedies in court. Follow-on applicants would be unable to seek declaratory judgment with respect to patents subject to notice by the patent proprietor prior to the commercial marketing of their products.[102]

In infringement litigation brought by the patent owner, the defendant may move to transfer the litigation to another district court that satisfies the requirements of jurisdiction and venue. In deciding whether to move the litigation to another court, a judge shall give greatest weight to the expeditious adjudication of the matter so that the biological product subject to the patent dispute may be brought to market as soon as possible, consistent with a fair and prompt resolution of the patent dispute.[103]

These bills also provide for data exclusivities that would be awarded to brand-name firms. Data exclusivities prevent the FDA from approving a follow-on version of the brand-name firm's product until a certain period of time had elapsed. H.R. 3962 (as passed by the House), H.R. 3590 (as passed by the Senate), S. 1679, H.R. 3200, and H.R. 1548 mandate that no application for a biosimilar product may be approved for 12 years from the reference product's licensure date. No application for a follow-on biologic may be filed for four years from the date the reference product was licensed. In addition, the first two bills state that exclusivity will not be awarded for (1) supplements to the reference product application; (2) the identification of new indications, routes of administration, dosing, or delivery; or (3) modifications to the structure of the biological product that do not result in a change in safety, purity, or potency.

In addition, H.R. 3590 and S. 1679 state that if a reference product has been designated an orphan drug, an application for a biosimilar or interchangeable product may not be filed until the later of: (1) the 7-year period of orphan drug exclusivity described in the FFDCA; or (2) the 12-year period established by this bill. Under H.R. 3962, H.R. 3200, and H.R. 1548, approval of pediatric applications would not become effective until 12 years and 6 months after approval of the reference product.

Pediatric exclusivity must be determined by the FDA no later than 9 months prior to the expiration of the initial marketing exclusivity period or no additional pediatric exclusivity will be awarded. H.R. 1548 further states the data exclusivity period is extended to 14 years if, during the 8-year period following licensing of the reference product, approval of a new indication which would provide significant improvement in treatment, diagnosis, or prevention.

Table 1. Proprietary Rights-Related Provisions of H.R. 3962 and H.R. 3590

Issue	H.R. 3962 Passed by House	H.R. 3590 Passed by Senate
Patent Infringement	H.R. 3962 mandates that within 30 days of the filing of a biosimilar application, the Secretary publish a notice of the application. The follow-on applicant also must provide the reference product owner with information concerning the biosimilar product and its manufacture. Within 60 days of the receipt of this information, the reference product sponsor must identify patents that it deems relevant to the biosimilar product. The reference product sponsor is required to update this list with any patents it subsequently acquires. In addition, "interested third parties" may inform the biosimilar applicant of their relevant patents at any time, and then be entitled to obtain information concern-ing the follow-on product and its production. The interested third party is then provided 90 days to identify further relevant patents; as with the reference product sponsor, the interested third party must update this list with subsequently acquired patents. Reference product sponsors and interested third parties must additionally state why they believe the biosimilar product would infringe their patents. The biosimilar applicant is then allowed 45 days to state either that it will not market its product until the relevant patents have expired, or alternatively provide its views that the patents are invalid or would not be infringed. In the latter case, the reference product owner of the interested third party may commence a patent infringement lawsuit against the biosimilar applicant within 60 days. [Sec. 2575]	H.R. 3590 requires that within 20 days after the Secretary publishes a notice that its application has been accepted for review, the biosimilar or interchangeable product applicant will provide the reference product sponsor with details concerning the product and its production. Within 60 days of the date of receipt of that information, the reference product sponsor is then required to identify patents that it deems relevant to the biosimilar or interchangeable product. The reference product sponsor is required to update this list with any patents it ubsequently acquires. The biosimilar or inter-changeable product applicant is then allowed 60 days to provide the reference product sponsor with a list of patents it deems relevant to the reference product. The biosimilar or interchangeable product applicant must also state either that it will not market its product until the relevant patents have expired, or alternatively provide its views that the patents are invalid or would not be infringed. In the latter case, the reference product sponsor or interested third party must provide the biosimilar or interchangeable product applicant with a response concerning the infringement and validity of those patents within 60 days. The parties are then required to engage in good faith negotiations to agree on which patents will be the subject of a patent infringement action. If those negotiations do not result in an agreement within 15 days, the parties will exchange lists of relevant patents to be litigated, with at least one patent identified by the reference product sponsor being subject to litigation. The reference product sponsor may then commence patent infringement litigation within 30 days. The biosimilar or interchangeable product applicant must notify the reference product sponsor at least 180 days before it commences

Table 1. (Continued)

Issue	H.R. 3962 Passed by House	H.R. 3590 Passed by Senate
		commercial marketing, and the reference product sponsor is allowed to seek a preliminary injunction at that time. Declaratory judgments are prohibited for either party unless the biosimilar or interchangeable product applicant fails to provide a product and patent application as required, at which time the reference product sponsor may seek a declaratory judgment. [Sec. 7002]
Data Exclusivity	An extra six months of data (market) exclusivity would be provided for a biologic drug if pediatric studies are conducted, with the exception of a biologic drug already on the market when the request for the extension on the marketed drug is made later than nine months before the expiration of the original exclusivity period. [Sec. 2575]	Similar provision; different wording. An extra six months of data (market) exclusivity would be provided for a new biologic drug if pediatric studies are conducted prior to FDA approval of the drug. An extra six months of data (market) exclusivity is provided for a biologic drug already on the market if pediatric studies are conducted and the request for the extension is made not less than nine months before the expiration of the original exclusivity period. [Sec. 7002]
Market Exclusivity	H.R. 3962 would allow for a period of exclusive marketing for the biological product that is the first to be established as interchangeable with the reference product. The provision also would provide a 12-year data exclusivity period (from the date on which the reference product was first approved) for the reference product. [Sec. 2575]	Same provision. [Sec. 7002]

Source: H.R. 3962, as passed by the House of Representatives on November 14, 2009, and H.R. 3590, as passed by the Senate on December 24, 2009.

For pediatric applications, approval of the follow-on biologic application would not become effective until 14 years and 6 months after approval of the reference product provided that the pediatric exclusivity is determined by FDA no later than 9 months prior to the end of the initial marketing exclusivity period.[104]

Under H.R. 1427 and S. 726, no application for a follow-on product may be approved for five years from the date of approval of the reference product if the innovator drug is considered a new major substance. Three years of exclusivity are granted to products that include a major substance that has been previously approved, provided that new clinical investigations essential to the approval of the application were undertaken. If the reference product has been subject to a supplemental application that the Secretary approves no later than one year before the expiration of this three- or five-year period, and the supplement relates to a change that provides a significant therapeutic advance, then the three- or five-year period is extended. If the gross sales of the reference product exceed one billion dollars, than the period of extension is three months; otherwise the period of extension is six months. [105] An additional six-month data exclusivity is awarded for brand-name biologics that are the subject of pediatric studies requested by the Secretary.[106]

Market exclusivity also would be awarded to the sponsor of a particular follow-on biologic that is the first to obtain FDA marketing approval. H.R. 3962 (as passed by the House), H.R. 3590 (as passed by the Senate), H.R. 3200, and S. 1679 allow a period of market exclusivity for the applicant that is the first to establish that its product is interchangeable with the reference product for any condition of use. The period of marketing exclusivity is the earlier of: (1) one year after the first commercial marketing of the first interchangeable biosimilar product to be approved as interchangeable with that reference product; (2) 18 months after either a final court judgment in patent infringement litigation under the Public Health Service (PHS) Act, as amended, or the dismissal of such litigation against the first applicant; (3) 42 months after the approval of the first interchangeable biosimilar biological product if patent litigation under the PHS Act, as amended, remains pending; or (4) 18 months after approval of the first interchangeable biosimilar product if the applicant has not been sued for patent infringement under the PHS Act, as amended.

H.R. 3962 also requires that reference product and biosimilar product sponsors file with the Assistant Attorney General and Federal Trade Commission copies of the text of any agreement they reach regarding the manufacture, marketing, or sale of either product. Agreements that solely concern purchase orders for raw materials, equipment and facility contracts, employment or consulting contracts, or packaging and labeling contracts are excluded. Failure to comply with this filing requirement may result in civil fines and other relief as the courts deem appropriate. With the concurrence of the Assistant Attorney General, the Federal Trade Commission may engage in rulemaking regarding the filing requirement.

H.R. 1427 and S. 726 permit a market exclusivity period for the first product that is determined to be interchangeable with the reference drug. This period would not expire until the earlier of (1) 180 days after first commercial marketing; (2) one year after a Court of Appeals decision on or dismissal of all patent infringement suits against the product; (3) 36 months after the date of approval if such patent litigation is on-going; or (4) one year after approval if there is no such litigation. [107] H.R. 1548 allows marketing exclusivity for the first follow-on biologic to establish that it is interchangeable with the reference product for one or more conditions of use. This period would be 24 months after the later of either the date of the first commercial marketing of the interchangeable drug or, if marketed before interchangeability is determined, the date the product is determined to be interchangeable.[108]

Innovation Issues

Patents have been particularly significant to the pharmaceutical industry. Creation of an expedited approval process for generic versions of traditional chemical drugs that have come off patent has been deemed very successful in bringing lower-cost versions of innovator products to the marketplace. However, while many biopharmaceuticals are now poised to lose, or recently have lost, patent protection, additional considerations not associated with typical chemical drugs may be brought into play in determining the success of efforts to bring similar, less expensive biotechnologies to the marketplace. Questions for policymakers remain as to whether or not an expedited approval process for biopharmaceuticals will result in a competitive market for these products similar to that created by the Hatch-Waxman Act for chemical drugs. The high costs associated with manufacturing biologics, the scale of additional clinical trials required for FDA marketing approval, and other marketing considerations may affect the availability of lower-cost versions of these products. The section below discusses the role of patents in innovation and the particular situation in the pharmaceutical industry. It then explores some of the unique issues associated with the manufacture and marketing of follow-on biologics to provide a context within which to assess various legislative options.

Patents and Innovation

Patent ownership is perceived to be an incentive to innovation, the basis for the technological advancement that contributes to economic growth. Article I, Section 8, Clause 8 of the U.S. Constitution states: "The Congress Shall Have Power... To promote the Progress of Science and useful Arts, by securing for limited Times to Authors and Inventors the exclusive Right to their respective Writings and Discoveries.... " Although not without question, patents are widely thought to encourage innovation by simultaneously protecting the inventor and fostering competition. They provide the inventor with a right to exclude others, temporarily, from use of the invention without compensation. Patents generally give the owner an exclusive right for 20 years (from date of filing) to further develop the idea, commercialize a product or process, and potentially realize a return on the initial investment. Concurrently, the process of obtaining a patent places the concept in the public arena. As a disclosure system, the patent can, and often does, stimulate other firms or individuals to invent "around" existing patents to provide for parallel technical developments or meet similar market needs.

The grant of a patent does not necessarily provide the owner with an affirmative right to market the patented invention. Pharmaceutical products are also subject to marketing approval by the Food and Drug Administration. Federal laws typically require that pharmaceutical manufacturers demonstrate that their products are safe and effective in order to bring these drugs to the marketplace. Issuance of a patent by the U.S. Patent and

Trademark Office and FDA marketing consent are distinct events that depend upon different criteria.

The patent system has dual policy goals—providing incentives for inventors to invent and encouraging inventors to disclose technical information.[109] Disclosure requirements are factors in achieving a balance between current and future innovation through the patent process, as are limitations on scope, novelty mandates, and nonobviousness considerations.[110] Many observers believe that patents give rise to an environment of competitiveness with multiple sources of innovation, which is viewed by some experts as the basis for technological progress. This may be important because, as Professors Robert Merges and Richard Nelson found in their studies, in a situation where only "a few organizations controlled the development of a technology, technical advance appeared sluggish."[111]

Not everyone agrees that the patent system is a particularly effective means to stimulate innovation. Critics argue that patents provide a monopoly that induces additional social costs. Some observers believe that the patent system encourages industry concentration and presents a barrier to entry in some markets.[112] Others believe that the patent system too frequently attracts speculators who prefer to acquire and enforce patents rather than engage in socially productive activity.[113] Still other commentators suggest that the patent system often converts pioneering inventors into technological suppressors, who use their patents to block subsequent improvements and thereby impede technological progress.[114]

Role of Patents in Pharmaceutical R&D

The utility of patents to companies varies among industrial sectors. Patents are perceived by pharmaceutical companies as critical to the drug industry. That may reflect the nature of R&D performed in this sector, where the resulting patents are more detailed in their claims and therefore easier to defend.[115] In contrast, one study found that in the aircraft and semiconductor industries patents are not the most successful mechanism for capturing the benefits of investments. Instead, lead time and the strength of the learning curve were determined to be more important.[116] Research undertaken by Professor Wesley Cohen and his colleagues demonstrated that patents were considered the most effective method to protect inventions in the drug industry when biotechnology is included.[117]

The high cost of drug development and the concomitant uncertainty associated with clinical trials necessary for marketing approval lends significance to patents in the pharmaceutical arena. Studies by Joseph DiMasi of Tufts University and others published in 2003 estimated that the capitalized cost of bringing a new drug (defined as a "new molecular entity" rather than a new formulation of an existing pharmaceutical product) to the point of marketing approval was $802 million (2000 dollars).[118] Additional research done by Federal Trade Commission analysts found the costs to be even higher; between $839 million and $868 million (2000 dollars).[119] At the same time, the total capitalized costs appear to be growing at an annual rate of 7.4% above general price inflation.[120]

A large portion of new drug costs (in terms of money and time) is associated with the size and breath of clinical trials necessary to obtain FDA marketing approval. According to a study supported by the Federal Reserve of Boston, only 10% of potential drug candidates reach the human trial phase and only a small portion of these actually reach the market. [121] In research presented at a conference sponsored by the Federal Reserve Bank of Dallas, Duke University's Henry Grabowski found that only 1% of drug compounds reach the human trial stage and 22% of those entering clinical trials receive FDA approval.[122] Professor Iain Cockburn notes that "as drug discovery became more science-intensive, ... it became not just more expensive but also more difficult to manage."[123] Furthermore, returns to new drug introductions vary widely and some experts have found that the median new drug does not bring in sufficient profits to cover the costs of bringing the product to the marketplace.[124] According to research by Professors Grabowski, John Vernon, and DiMasi, only 34% of new drugs (new chemical entities) introduced generated profits that equaled the industry average R&D cost.[125]

The degree to which industry perceives patents as effective has been characterized as "positively correlated with the increase in duplication costs and time associated with patents."[126] In certain industries, patents significantly raise the costs incurred by nonpatent holders wishing to use the idea or invent around the patent—an estimated 40% in the pharmaceutical sector, 30% for major new chemical products, and 25% for typical chemical goods—and are thus viewed as significant. However, in other industries, patents have much smaller impact on the costs associated with imitation (e.g., in the 7%-1 5% range for electronics), and may be considered less successful in protecting resource investments.[127]

The significant costs of pharmaceutical R&D, coupled with the uncertainty of the clinical trial process, lend consequence to patents in this area because "the disparity between the investments of innovators and those of imitators is particularly large in pharmaceuticals—almost as large as when software pirates simply copy the diskettes of an innovator."[128] While the capitalized cost of developing a new drug to the point of market approval is estimated at over $800 million, it takes only between $1 million and $2 million to obtain approval for a generic version of the chemically synthesized pharmaceutical. [129] This difference is a result of the costs associated with clinical trials needed to demonstrate the safety and efficacy of a new drug, data that could be utilized by generic companies if not protected by a patent.[130]

Clinical Trials

The scale and extent of clinical trials necessary to approve follow-on biologics is expected to factor into whether or not this industry will provide the cost savings needed to be viable.[131] The varied characteristics of individual biologic products may make it likely that regulatory and developmental requirements for follow-on products will need to reflect each individual situation.[132] Innovator and generic manufacturers appear to agree that "Unlike small-molecule copycats, for biogenerics, the nature and extent of the data needed will also depend very much on the product involved: regulatory guidelines must be defined product by product."[133]

Generic biotechnology manufacturers assert that extensive new clinical trials would not be necessary in an expedited approval process but should be considered on a case-by-case basis, as discussed previously. Innovator companies dispute this. The nature of biologics has resulted in longer mean clinical development time for these products when compared with traditional drugs.[134] In addition, the number of clinical trials necessary to file a new drug application has doubled since 1980 and the number of participants in these trials has tripled.[135] If additional clinical trials are necessary to demonstrate "sameness," effectiveness, and safety, estimates are that it may take twice the time to develop a follow-on biopharmaceutical than a chemical generic with a cost that some expect to be 8-100 times higher than that associated with a traditional generic product.[136] Phase III trials are the most expensive of the required trials and any additional requirements for follow-on biologics likely would increase the cost to the public.[137]

Manufacturing Considerations

As discussed above, biotechnology drugs are characterized by their manufacturing process such that:

> The manufacturing process for each biologic defines, to a significant extent, the product because biologics are based on living cells or organisms whose metabolisms are inherently variable. Moreover, apparently small differences between manufacturing processes can cause significant differences in the clinical properties of the resulting products.[138]

This insures that the manufacture of biologics will tend to be significantly more expensive than traditional chemically synthesized drugs.[139] The FDA is required to inspect the manufacturing facilities and processes involved in the production of biologics: "Unlike small-molecule manufacturing, biomanufacturers get approval for both the drug and the process used to make it, and that approval can take years."[140] Therefore, these facilities must be built and operational prior to the FDA approval process. According to FDA guidelines, "Issuance of a biologics license is a determination that the product, *the manufacturing process, and the manufacturing facilities* [emphasis added] meet applicable requirements to ensure the continued safety, purity and potency of the product."[141]

It has been estimated that each large U.S.-based biologic "manufacturing facility costs between $200 and $400 million to build, and takes four years before gaining approval by the US Food and Drug Administration."[142] In addition, the cost of materials to manufacture biologics may be 20 to 100 times more than chemical drugs.[143] The production process for biologics typically takes longer than traditional drugs and may take eight to nine months.[144]

Altering the manufacturing process in any way tends to require that validation be repeated.[145] One commentator stated: "It's hard to predict how process variations will change a product's safety or effectiveness."[146] This is a result of the incidence of impurities arising from changes in the method of production and the increased opportunity of adverse immune reactions.[147] It is difficult to find and identify impurities in biologics as, to date, simple tests do not exist. Thus, there are often additional costs associated with preventing impurities from entering into the production process.[148] Some experts argue that there is also a need for additional clinical (human) trials to insure that any changes to the production process do not result in impurities that are harmful.[149] However, generic manufacturers assert that they can maintain high standards in the manufacturing process to insure similar products that are safe and effective.[150]

Marketing Concerns

Several commentators have suggested that marketing costs associated with follow-on biologics will be higher than with traditional generics because of the need to convince doctors that these products generate similar results.[151] If the follow-on biopharmaceutical cannot be termed equivalent to the brand name drug, doctors and pharmacists may not be able to readily substitute:

> Marketing and patient support are more important for biosimilars, favouring companies with significant financial resources and who have had experience in marketing branded products. The generics market has historically used prices to secure market share, so it is important for biosimilar developers to understand and act on these factors. Early-stage success in the biosimilars market, however, is more dependent on the speed to market and successful marketing strategies.[152]

The greater the number of generic alternatives, the less the cost. However, biologics may not generate multiple follow-on products for the same brand name biopharmaceutical because of the higher costs associated with bringing these drugs to the marketplace. Price differentials associated with follow-on products may not be as great as with other generics because of the large initial costs related to establishing manufacturing facilities and performing any additional clinical studies necessary for FDA approval. Therefore, the makers of follow-on products would be expected to charge higher prices and generate more profits than the typical generic firm.[153] In addition, "Financial and scientific barriers might prevent the cutthroat price wars fought in the traditional generic market."[154] A study by Kalorama Information (*The Market for Generic Biologics: Issues, Trends, and Market Potential*, June 1, 2005) estimated that follow-on products will sell for only 10-20% less than the brand name biologic, not the 40-80% reduction in price generally seen with chemical drug generics.[155] A Merrill Lynch analysis[156] estimated prices 20%- 30% below the brand biologic for the first biosimilar to be marketed while a report by Citizens Against Government Waste[157] estimated savings of 10%-25% over the brand biologic price in the first year and 25%-47% by the fifth year after introduction of a follow-on drug.

Additional Observations

If alternative mechanisms for accelerated approval of follow-on biologics are legislated, many experts argue that the cost savings will not be as substantial as those generated by typical generic drugs. High manufacturing costs, the need for additional safety and efficacy trials to test these products, and augmented marketing efforts directed at doctors and patients to encourage the use of similar, but not identical drugs, are expected to add to the prices associated with the follow-on product. The differences between biologics and other drugs

> in turn lead to important differences in the economics of discovery, development, manufacturing, and distribution for drugs and biologics. Consequently, this could lead to different economic outcomes in terms of average prices, number of competitors, returns on spending for research and development (R&D), and other market measures.[158]

Some commentators are concerned that an expedited generic approval process similar to that established in the Hatch-Waxman Act may raise issues associated with the affect of patent challenges on biotechnology companies, many of which do not make a profit. How might this impact upon innovation in this sector? It has been argued that "[n]inety percent of biotech companies are surviving on venture capital, do not yet have a single product on the market, and are working hard to move products through preclinical discovery and chemistry to clinical investigation and then through FDA approval."[159] Often, a firm's intellectual property is its primary asset, particularly through the drug development stage, and typically is utilized to raise funds for additional R&D. Thus, several experts maintain that defending patents may divert support from on-going innovation, especially in small companies that make up a significant portion of the biotechnology sector.

Other experts argue that "Opening up biotech drugs to the prospect of generic competition *after* [emphasis added] patents expire may even spark innovation—forcing biotech companies to come up with improved versions of existing drugs that perhaps require less-frequent dosing, have fewer side effects or hang around the body longer, making them more effective."[160] However, the ability of brand name companies to bring out improved versions of their initial biologic may dampen development of follow-on products.[161] Because biotechnology is advancing so quickly, improvements in existing products may dissuade generic firms from making follow-on products that require large investments in manufacturing plants and clinical trials.[162] Patients may just switch to the next generation brand name drug.

A question remains whether or not multiple companies will invest the time and money necessary to develop follow-on biologics:

> The scope of the requirements means that, in organizational terms, the development of biogenerics demands a culture and mentality closer to that of proprietary pharmaceutical developers than to that of conventional generics firms.... If regulatory demands for biogenerics prove exacting, then [established pharmaceutical] companies are probably better off developing entirely novel biotech products instead. The regulatory requirements for these products are usually much clearer, and in most cases, companies will stand to make far better commercial returns by taking this road.[163]

On the other hand,

> One reason generics companies are dropping out of the race is the sheer scope of demands these projects require.... Generic companies also tend to be much less tolerant of delays and setbacks.... Given these circumstances, most generics companies will have a hard time coming to grips with the demands of biogenerics.[164]

If fewer companies chose to make follow-on products, there would be less competition in the marketplace resulting in reduced cost savings. This raises the issue of whether there are other, or additional mechanisms to encourage firms to produce lower cost follow-on biologics.

In accessing any potential legislative activity in this area, it might be important to consider how to facilitate follow-on products after patent expiration while continuing to encourage innovation in the brand name biopharmaceutical sector. The Hatch-Waxman Act attempted to build just such a balance between the introduction of widely available generic drugs with adequate incentives for investment in the development of new pharmaceuticals. Many policy and industry experts agree that the Hatch-Waxman Act has had a significant effect on the availability of generic substitutes for brand name drugs while investment in the development of new pharmaceuticals has continued. At issue is whether or not it is desirable to pursue, and possible to achieve, a similar balance of interests in the biopharmaceutical industry.

Acknowledgments

This chapter was funded in part by a grant from the John D. and Catherine T. MacArthur Foundation.

End Notes

[1] The term "biologic" has been described as "poorly defined," and its precise parameters are themselves subject to debate. *See* David M. Dudzinski, "Reflections on Historical, Scientific, and Legal Issues Relevant to Designing Approval Pathways for Generic Versions of Recombinant Protein-Based Therapeutics and Monoclonal Antibodies," 60 *Food and Drug Law Journal* (2005), 143. The Public Health Service defines the term "biological product" to mean "a virus, therapeutic serum, toxin, antitoxin, vaccine, blood, blood component or derivative, allergenic product, or analogous product, or arsphenamine or derivative of arsphenamine (or any other trivalent organic arsenic compound), applicable to the prevention, treatment, or cure of a disease or condition of human beings." 42 U.S.C. § 262(i) (2006). Biologics are also sometimes termed "biotechnology drugs" or "biopharmaceuticals." Dudzinski at 143.

[2] Biotechnology Industry Organization, available at http://www.bio.org and Ernst & Young, *Beyond Borders, Global Biotechnology Report 2007*, 25.

[3] PriceWaterhouseCoopers, *Biotech, Lifting Big Pharma 's prospects with biologics*, May 2009, 7, available at http://www.pwc.com/extweb/industry.nsf/docid/5C4FDB15B185FE39802575C400271BD9/$File/biotech-final.pdf.

[4] Kerry A. Dolan, "Biology Rising," *Forbes.com*, May 12, 2006, available at http://www.forbes.com/2006/05/12/merck-pfizer-amgen-cz_kd_05 12biologics_print.html.

[5] Ernst & Young, *Beyond Borders, Global Biotechnology Report 2008*, http://www.ey.com/Global/assets International/Industry_Biotechnology_Beyond_Borders_2008/$file/Biotechnology_Beyond_Borders_2008.pdf.

[6] "Biogenerics: A Difficult Birth?" *IMS Global Insights*, available at http://www.imshealth.com.

[7] Nicole Gray, "Harbingers of Change," *Pharmaceutical Executive*, May 2005, available at http://www.imshealth.com; Nicole Gray, "Keeping Pace with the Evolving Pharmaceutical Business Model," *Pharmaceutical Executive*, May 2006, available at http://www.imshealth.com.

[8] Ramsey Baghdadi, "Biogenerics Are Happening: Slowly, Product-By-Product," *The RPM Report*, January 2006, available at http://www.theRPMreport.com.

[9] Aaron Smith, "Barr's Risky $2.5 Billion Bid for Biogenerics," *CNNMoney .com*, September 15, 2006. *See also* Research and Markets, "The Biogenerics Market Outlook: An Analysis of Market Dynamics, Growth Drivers and Leading Players," *Business Wire*, September 12, 2005, available at http://www.findarticles.com/p/articles/mi_mOEIN/ is_2005_Sept_12/ai_n 153 82946/print.

[10] Parexel Consulting, *Follow-on Biologics*, availabe at http://www.parexelconsulting.com/services/follow-on-biologics.html.

[11] FierceBiotech, *The Top 10 Biosimilar Players: Positioning, performance and SWOT analyses*, Aprile 20, 2009, available at http://www.fiercebiotech.com/research/top-10-biosimilar-players-positioning-performance-and-swotanalyses.

[12] PriceWaterhouseCoopers, *Biotech: Lifting Big Pharma 's prospects with biologics*, 2009, p. 9, available at http://www.pwc.com/en_GX/gx/pharma-life-sciences/pdf/biotech-final.pdf.

[13] "2008 Top 200 branded drugs by retail dollars," *Drug Topics*, 2009 available at http://www.drugtopics.com.

[14] Ed Lamb, Patent Expirations Could Mean Biologics Competition, Pharmacy Times, March 1, 2008, available at http://www.pharmacytimes.com/issue/pharmacy/2008/2008-03/2008-03-8430.

[15] "2008 Top 200 branded drugs by retail dollars," *op. cit.*

[16] *Id.*

[17] Thomas Gryta, "Generic Biologics Face Hurdles, The Wall Street Journal, March 4, 2009 available at http://online.wsj.com/article/SB123614292238326905.html.

[18] *See* Dudzinski, supra.

[19] P.L. 84-417, 98 Stat. 1585 (1984).

[20] *See* 21 U.S.C. § 355(b)(2) (2006) (with respect to § 505(b)(2) applications); *id.* § 355(j)(1) (with respect to ANDAs).

[21] *See* John R. Thomas, *Pharmaceutical Patent Law* (Bureau of National Affairs, 2005) and CRS Report RL30756, *Patent Law and Its Application to the Pharmaceutical Industry: An Examination of the Drug Price Competition and Patent Term Restoration Act of 1984 ("The Hatch-Waxman Act ")*, by Wendy H. Schacht and John R. Thomas.

[22] *See* Melissa R. Leuenberger-Fisher, "The Road to Follow On Biologics: Are We There Yet?," 23 *Biotechnology Law Report* (August 2004), 389.

[23] *See* Dawn Willow, "The Regulation of Biologic Medicine: Innovators' Rights and Access to Healthcare," 6 *Chicago- Kent Journal of Intellectual Property* (2006), 32.

[24] *Id.*

[25] *See* Dudzinski, *supra* (noting such concerns).

[26] *See* Narinder S. Banait, *Follow-on Biological Products—Legal Issues*, 2005, available at http://www.fenwick.com/ docstore/Publications/IP/follow-on.pdf.

[27] *See, e.g,* J.E.M. Ag Supply v. Pioneer Hi-Bred Int'l, Inc., 534 U.S. 124, 130 (2001) (explaining patent law principle that "products of nature" are not eligible for patenting, but that patents may be available for "human-made inventions" resulting from the use of biotechnology).

[28] *See* Thomas, *supra* note 11.

[29] P.L. 84-417, 98 Stat. 1585 (1984).

[30] For additional information see CRS Report RL3 0756, *Patent Law and Its Application to the Pharmaceutical Industry: An Examination of the Drug Price Competition and Patent Term Restoration Act of 1984 ("The Hatch- Waxman Act")*, by Wendy H. Schacht and John R. Thomas, and CRS Report RL32377, *The Hatch-Waxman Act: Legislative Changes Affecting Pharmaceutical Patents*, by Wendy H. Schacht and John R. Thomas.

[31] *Id.*

[32] *See* Apotex Inc. v. FDA, 393 F.3d 210, 212 (D.C. Cir. 2004).

[33] *See* Alfred B. Engelberg, "Special Patent Provisions for Pharmaceuticals: Have They Outlived Their Usefulness?" 39 *IDEA: J. L. & TECH.* 389, 396 (1999).

[34] *See* James J. Wheaton, "Generic Competition and Pharmaceutical Innovation: The Drug Price Competition and Patent Term Restoration Act of 1984," 34 *CATH. UNIV. L. REV.* 433, 439 (1986).

[35] *See, e.g.,* Justina A. Molzon, "The Generic Drug Approval Process," 5 *J. PHARM. & L.* 275, 276 (1996) ("The Act streamlined the approval process by eliminating the need for [generic drug] sponsors to repeat duplicative, unnecessary, expensive and ethically questionable clinical and animal research to demonstrate the safety and efficacy of the drug product.").

[36] *See* Jonathon M. Lave, "Responding to Patent Litigation Settlements: Does the FTC Have It Right Yet?" 64 *U. PITT. L. REV.* 201, 202 (2002) ("Hatch-Waxman has also increased the generic drug share of prescription drug volume by almost 130% since its enactment in 1984. Indeed, nearly 100% of the top selling drugs with expired patents have generic versions available today versus only 35% in 1983.").

[37] 21 U.S.C. § 355(j)(1) (2006).

[38] *See, e.g.,* Sarah E. Eurek, "Hatch-Waxman Reform and Accelerated Entry of Generic Drugs: Is Faster Necessarily Better?" 2003 *DUKE L. & TECH. REV.* 18 (August 13, 2003).

[39] 21 U.S.C. § 355(b)(2) (2004). The name of this application refers to its section number within the Hatch-Waxman Act itself. This provision has been codified at 21 U.S.C. § 355(b)(2), where it may be more conveniently located.

[40] U.S. Department of Health and Human Services, Food and Drug Administration, Center for Drug Evaluation and Research (CDER), *Guidance for Industry: Applications Covered by Section 505(b) (2)* at 2-3 (October 1999) [hereinafter "*Section 505 (b) (2) Guidance*"].

[41] *Id.*

[42] *Id.*

[43] *See* Pfizer, Inc. v. Dr. Reddy's, 359 F.3d 1361, 69 USPQ2d 2016 (Fed. Cir. 2004).

[44] 37 C.F.R. § 314.93(b) (2004).

[45] *Section 505(b) (2) Guidance* at 5.

[46] *See* Bruce N. Kuhlik, "The Assault on Pharmaceutical Intellectual Property," 71 *U. CHI. L. REV.* 93, 103-04 (2004).

[47] *Section 505(b) (2) Guidance* at 3.

[48] The FDC Act defines the term "drug" to include "articles intended for use in the diagnosis, cure, mitigation, treatment, or prevention of disease in man or other animals.... " 21 U.S.C. § 321(g)(1)(B).

[49] P.L. 78-410, 58 Stat. 682 (1944).

[50] U.S. Food and Drug Administration, Center for Drug Evaluation and Research, "Frequently Asked Questions About Therapeutic Biological Products"(available at http://www.fda.gov/cder/biologics/qa.htm).

[51] Further discussion of this issue may be found at Willow, *supra.*

[52] Gary C. Messplay and Colleen Heisey, "Follow-On Biologics: The Evolving Regulatory Landscape," *BioExecutive International*, May 2006, 43.

[53] Stephen B. Judlowe and Brian P. Murphy, *Proposed Legislation for Follow-On Biologic Pharmaceuticals in the US*, Morgan, Lewis & Backius, LLP, New York available at http://www.buildingipvalu e.com/05_NA/135_138.htm.

[54] Debra Weintraub, "Next Generation of Biopharmaceuticals," *Journal of Managed Care Medicine*, vol. 9, no. 1, 2006, available at http://www.namcp.com/Journals/JMCM/Articles/ Next%20Generation%20of%20Bio pharmaceuticals.pdf.

[55] Henry Grabowski, Iain Cocburn, and Genia Long, "The Market For Follow-On Biologics: How Will It Evolve?," *Health Affairs,* September/October 2006, 1292.

[56] Judlowe & Murphy, *supra* note 44.

[57] Bruce S. Manheim Jr., Patricia Granahan, and Kenneth J. Dow, "'Follow-On Biologics': Ensuring Continued Innovation In The Biotechnology Industry," *Health Affairs*, March/April 2006, 397.

[58] *Id.*

[59] *See* Weintraub, *supra* note 45.

[60] Judlowe & Murphy, *supra*.

[61] *See* Christopher Webster et al., "Biologics: Can There Be Abbreviated Applications, Generics, or Follow-On Products?," *International BioPharm* (July 1, 2003) (available at http://www.biopharm-mag.com/biopharm/article/ articleDetail.jsp?id=73785).

[62] Judlowe & Murphy, *supra*.

[63] Generic Pharmaceutical Association, *Generic Biopharmaceuticals*, available at http://www.gphaonline.org.

[64] Selena Class, "Biogenerics: Waiting for the Green Light," *IMS Global Insights*, October 28, 2004 (available at http://www.imshealth.com).

[65] *Id.*

[66] Narinder S. Banait, "Follow-on Biological Products—Legal Issues" (2005) (available at http:// www.fenwick.com).

[67] 21 U.S.C. § 355(b)(2) (2006).

[68] Biotechnology Industry Organization, *Follow-on Biotechnology Products* (available at http://www.bio.org/ healthcare/followon/).

[69] For additional information, see CRS Report RL34045 , *FDA Regulation of Follow-On Biologics*, by Judith A. Johnson.

[70] H.R. 1548 at § 101(a)

[71] H.R. 1427 at § 3(a), S. 726 at § 3(a)

[72] *See* CRS Report RL3 0648, *An Examination of the Issues Surrounding Biotechnology Patenting and Its Effect Upon Entrepreneurial Companies*, by John R. Thomas.

[73] *See* 35 U.S.C. §§ 101, 102, 103 (2006).

[74] 35 U.S.C. § 271(2006).

[75] 35 U.S.C. §§ 101, 102, 103 (2006).

[76] *See* Brenner v. Manson, 383 U.S. 519, 86 S.Ct. 1033, 16 L.Ed.2d 69 (1966).

[77] 35 U.S.C. § 102 (2006).

[78] 35 U.S.C. § 103(a) (2006).

[79] 35 U.S.C. § 111 (2006).

[80] 35 U.S.C. § 112 (2006).

[81] 35 U.S.C. § 112 (2006).

[82] 35 U.S.C. § 151 (2006).

[83] 35 U.S.C. § 154 (2006).

[84] 35 U.S.C. § 156 (2006).

[85] 35 U.S.C. § 271(a) (2006).

[86] 35 U.S.C. § 281 (2006).

[87] 35 U.S.C. § 282 (2006).

[88] 473 U.S. 303 (1980).

[89] *Id.* at 309-10.

[90] *See* Diamond v. Diehr, 450 U.S. 175, 185 (1980) ("Excluded from such patent protection are laws of nature, natural phenomena, and abstract ideas.").

[91] U.S. Patent No. 5,955,422 (claiming in part "[a] pharmaceutical composition comprising a therapeutically effective amount of human erythropoietin and a pharmaceutically acceptable diluent, adjuvant or carrier, wherein said erythropoietin is purified from mammalian cells grown in culture."). *See also* Amgen, Inc. v. Hoechst Marion Roussel, Inc., 457 F.3d 1293 (Fed. Cir. 2006).

[92] *See* Alix Weisfeld, "How Much Intellectual Property Protection Do the Newest (and Coolest) Biotechnologies Get Internationally?," 6 *Chicago Journal of International Law* (2006), 833.

[93] *See* Dennis J. Karjala, "Biotech Patents and Indigenous Peoples," 7 *Minnesota Journal of Law, Science, and Technology* (2006), 483.

[94] P.L. 84-417, 98 Stat. 1585 (1984).

[95] A more detailed discussion of the Hatch-Waxman Act is found at CRS Report RL30756, *Patent Law and Its Application to the Pharmaceutical Industry: An Examination of the Drug Price Competition and Patent Term Restoration Act of 1984 ("The Hatch-Waxman Act")*, by Wendy H. Schacht and John R. Thomas.

[96] *See* Patent Fairness Act of 1999: Hearing on H.R. 1598 Before the House Comm. on the Judiciary, 106th Cong. 94 (1999) (statement of Hon. Henry A. Waxman, Member, House Comm. on Government Reform) (available at http://commdocs.house.gov/committees/judiciary

[97] The term "data exclusivity" refers to a period of time during which an applicant for FDA marketing approval may not rely upon the preclinical, clinical, or other data developed and submitted by another firm. Data exclusivity does not preclude the FDA from approving an application for marketing approval when the applicant has itself conducted, or has a right of access to, preclinical, clinical, or other data generated by another company. Some sources refer to data exclusivity as "marketing exclusivity" or "data protection." *See* Thomas, supra, at 349.

[98] *See* Terry Mahn and Margo Furman, *The Role of Patent and Non-Patent Exclusivity Under the Hatch-Waxman Act* (2005), 19-20 (available at http://www.fr.com/practice/pdf/05-16-05%20_PatentandNon-PatentExclusivity.pdf).

[99] All quotes in this section, unless otherwise noted, are taken from responses to a questionnaire sent to 35 stakeholders interested in the effort to craft an accelerated approval pathway by the House Committee on Energy and Commerce dated April 3, 2008 and posted on the Committee website at http://energycommerce.house.gov/Press_110/ 040308.FOB.Responses.shtml

[100] H.R. 1548 at § 101(a)

[101] H.R. 1427 at § 3(a) and S. 726 at § 3(a).

[102] A declaratory judgment action is provided by a "federal or state law permitting parties to bring an action to determine their legal rights and positions regarding a controversy not yet ripe for adjudication.... " *Black's Law Dictionary* (8th ed. 2004). In the context of patents, an action for declaratory judgment potentially would allow a biosimilar applicant to "clear the air by suing for a judgment that would settle the conflict of interests" with the owner of the reference product. *Arrowhead Indus. Water, Inc. v. Ecolochem, Inc.*, 846 F.2d 731, 735 (Fed. Cir. 1988).

[103] H.R. 1427 at § 3(b) and S. 726 at § 3(b).

[104] H.R. 1548 at § 101(a).

[105] H.R. 1427 at § 3(a) and S. 726 at § 3(a).

[106] H.R. 1427 at §4 and S. 726 at § 4.

[107] H.R. 1427 at § 3(a) and S. 726 at § 3(a)

[108] H.R. 1548 at § 101(a)

[109] Robert P. Merges, "Commercial Success and Patent Standards: Economic Perspectives on Innovation," *California Law Review*, July 1988, 876.

[110] Keneth W. Dam, "The Economic Underpinnings of Patent Law," *Journal of Legal Studies*, January 1994, 266-267. Scope is determined by the breath of the claims made in a patent. Claims are the technical descriptions associated with the invention. In order for an idea to receive a patent, the law requires that it be "new, useful [novel], and nonobvious to a person of ordinary skill in the art to which the invention pertains."

[111] Robert P. Merges and Richard R. Nelson, "On the Complex Economics of Patent Scope," *Columbia Law Review*, May 1990, 908.

[112] *See* John R. Thomas, "Collusion and Collective Action in the Patent System: A Proposal for Patent Bounties," *University of Illinois Law Review*, 2001, 305.

[113] *Id.*

[114] *On the Complex Economics of Patent Scope*, 839.

[115] *Id.*, 255 and 257. *See also*: Edwin Mansfield, "Intellectual Property Rights, Technological Change, and Economic Growth," in *Intellectual Property Rights and Capital Formation in the Next Decade*, eds. Charles E. Walker and Mark A. Bloomfield (New York: University Press of American, 1988), 12 and 13.

[116] Richard C. Levin, Alvin K. Klevorick, Richard R. Nelson, and Sidney G. Winter. "Appropriating the Returns for Industrial Research and Development," *Brookings Papers on Economic Activity*, 1987, in *The Economics of Technical Change*, eds. Edwin Mansfield and Elizabeth Mansfield (Vermont, Edward Elgar Publishing Co., 1993), 253.

[117] Wesley M. Cohen, Richard R. Nelson, and John P. Walsh, *Protecting Their Intellectual Assets: Appropriability Conditions and Why U.S. Manufacturing Firms Patent (or Not)*, NBER Working Paper 7552, Cambridge, National Bureau of Economic Research, February 2000, available at http://www.nber.org/papers/w7552.

[118] Joseph A. DiMasi, Ronald W. Hansen, and Henry G. Grabowski. "The Price of Innovation: New Estimates of Drug Development Costs," 22 *Journal of Health Economics,* 2003. Capitalized cost includes the "time cost" associated with an investment and the cost of testing drug products that fail.

[119] Christopher P. Adams and Van V. Brantner, *Estimating the Costs of New Drug Development: Is it Really $802m?*, Federal Trade Commission, December 2004, available at http://media

[120] *The Price of Innovation: New Estimates of Drug Development Costs*, 180.

[121] Carrie Conway, "The Pros and Cons of Pharmaceutical Patents," *Regional Review*, Federal Reserve Bank of Boston, March, 2003, available at http://www.findarticles.com.

[122] Henry G. Grabowski, "Patents, Innovation, and Access to New Pharmaceuticals," *Journal of International Economic Law,* 2002, 851.

[123] Iain Cockburn, "The Changing Structure of the Pharmaceutical Industry," *Health Affairs*, January/February 2004, 15.

[124] Henry G. Grabowski, "Patents and New Product Development in the Pharmaceutical and Biotechnology Industries," *Science and Cents: Exploring the Economics of Biotechnology*, Proceedings of a 2002 Conference, Federal Reserve Bank of Dallas, 95-96 available at http://www.dallasfed.org/research/pubs/science/grabowski.pdf; and Henry Grabowski, John Vernon, and Joseph A. DiMasi, "Returns on Research and Development for 1990s New Drug Introductions," 20 *Pharmacoeconomics*, 2002.

[125] Henry Grabowski, John Vernon, and Joseph A. DiMasi, "Returns on Research and Development for 1990s New Drug Introductions," 20 *Pharmacoeconomics*, 2002, 23.

[126] *Appropriating the Returns for Industrial Research and Development,* 269.

[127] Edwin Mansfield, Mark Schwartz, and Samuel Wagner. "Imitation Costs and Patents: An Empirical Study," *The Economic Journal*, December 1981, in *The Economics of Technical Change, 270. See also: Appropriating the Returns for Industrial Research and Development.*

[128] Federic M. Scherer, "The Economics of Human Gene Patents," 77 *Academic Medicine*, December 2002, 1352.

[129] *Patents, Innovation, and Access to New Pharmaceuticals*, 852.

[130] *The Economics of Human Gene Patents*, 1352.

[131] "Delay in U.S. Regulatory Approval Significantly Lowers Forecast for BioGenerics Market to $2.3 Billion," *Business Wire*, November 22, 2005.

[132] John Ansell, "Biogenerics Part I: Set to Make Real Inroads or Not?," *PharmaWeek*, January 26, 2006, available at http://www.pharmaweek.com?Exclusive_Content/1_26.asp.

[133] *Id.*

[134] *The Market For Follow-On Biologics: How Will It Evolve?*, 1293.

[135] Gregory J. Glover, "The Influence of Market Exclusivity on Drug Availability and Medical Innovations," *The AAPS Journal*, August 3, 2007, E3 13.

[136] IMS Health, "Biogenerics: A Difficult Birth?," May 18, 2004 available at http://www.imshealth. com/web 0,3 148,64576068_63872702_70261000_71026746,00.html.

[137] Ernst & Young, "Coming of Age," *Beyond Boarders*, 2005, available at http://www.ey.com/beyondboarders.

[138] *Biologics: Can There Be Abbreviated Applications, Generics, or Follow-On Products*?

[139] *The Market For Follow-On Biologics: How Will It Evolve?*, 1293; and *The Long and Winding Road to Biologic Follow-ons*, 24.

[140] Alison McCook, "Manufacturing on a Grand Scale," *The Scientist*, February 14, 2005, available at http://www.thescientist.com.

[141] U.S. Food and Drug Administration, *Frequently Asked Questions About Therapeutic Biological Products*, July 26, 2006, available at http://www.fda.gov/cder/biologics/qa.htm.

[142] *Manufacturing on a Grand Scale.*

[143] *The Market For Follow-On Biologics: How Will It Evolve?,* 1293.

[144] *The Long and Winding Road to Biologic Follow-ons*, 24.

[145] *Manufacturing on a Grand Scale.*

[146] William Alpert, "Biotech's Next Challenge," *SmartMoney .com*, May 22, 2006, available at http://www.smartmoney.com/barrons/index.cfm?story=20060522.

[147] Joshua W. Devine, Richard R. Cline, and Joel F. Farley, "Follow-on Biologics: Competition in the Biopharmaceutical Marketplace," *Journal of the American Pharmacists Association*, March/April 2006, 194.

[148] Gurdeep Singh Shah, "The Current Market for Generic Biologics," International Biopharmaceutical Association, June 2006, available at http://www.ibpassociation.org/IBPA_articles/jun2006issue/ The _Current _Market _for _Generic _Biologics.htm.

[149] *Biologics: Can There Be Abbreviated Applications, Generics, or Follow-On Products?*

[150] *Proposed Legislation for Follow-On Biologic Pharmaceuticals in the US.*

[151] *The Long and Winding Road to Biologic Follow-ons*, 24.

[152] Mark J. Belsey, Laura M. Harris, Romita R. Das, and Joanna Chertkow, "Biosimilars: Initial Excitement Gives Way to Reality," *Nature Reviews Drug Discovery*, July 2006, available at http://www.nature.com /nrd/journal/v5/n7/full/ nrd2093.html.

[153] William Alpert, "Biotech's Next Challenge," *SmartMoney .com*, May 22, 2006, available at http://smartmoney.com/ print/index.cfm?printcontent=/barrons/index.cfmstory=20060522.

[154] *Id.*

[155] Susan J. Ainsworth, "Biopharmaceuticals," *Chemical and Engineering News*, June 6, 2005, 21-29.

[156] Merrill Lynch, *Biogenerics: Big Opportunities, Small Threat*, September 6, 2006.

[157] Everett Ehrlich and Elizabeth L. Wright, Biogenerics: What They Are, Why They Are Important, and Their Economic Value to Taxpayers and Consumers, Citizens Against Government Waste, May 2, 2007.

[158] *The Market For Follow-On Biologics: How Will It Evolve?*, 1292.

[159] *See* Weintraub, *supra.*

[160] Scott Gottlieb, *Biologics Warfare*, May 6, 2005, available at http://www.aei.org/publication22467.

[161] *Biopharmaceuticals.*

[162] *Id.*

[163] *Biogenerics Part I: Set to Make Real Inroads or Not?*

[164] *Id*

In: Follow-On Biologics: Background and Issues
Editor: Daniel J. Gutiérrez
ISBN: 978-1-61122-076-6
© 2011 Nova Science Publishers, Inc.

Chapter 4

Emerging Health Care Issues: Follow-On Biologic Drug Competition

Federal Trade Commission Report

Executive Summary

Over the past three decades, the lure of patent protection, coupled with the ability to price at market rates, has spurred pioneer drug manufacturers to develop new therapeutic drugs known as biologics. These innovations have improved medical treatments, reduced suffering, and saved the lives of many Americans. Biologic drugs are protein-based and derived from living matter or manufactured in living cells using recombinant DNA biotechnologies. The therapeutic proteins that form the basis of these biologic drugs are far more complex and much larger than the chemically synthesized, small molecules that form the basis of most pharmaceutical products.

Biologic drug innovations, however, are expensive. As examples, annual treatment for breast cancer with the biologic drug Herceptin can cost $48,000 and the annual treatment for rheumatoid arthritis with Remicade can cost approximately $20,000. Indeed, in 2007, Americans spent $286.5 billion for prescription drugs, $40.3 billion of which was for biologic drugs.[1]

Questions have arisen whether the price of biologics might be reduced by competition if there were a statutory process to encourage "follow-on biologics" ("FOBs") to enter and compete with pioneer biologics once a pioneer drug's patents have expired. The obvious model for such a statute is the Hatch-Waxman Act, which Congress enacted in 1984 to allow the Food and Drug Administration ("FDA") to approve the sale of generic versions of branded drugs, among other things.[2] The Hatch-Waxman Act does not apply to biologics, which the FDA approves pursuant to the Public Health Safety ("PHS") Act. Rather, Hatch-Waxman applies only to drugs regulated under the Federal Food Drug and Cosmetic Act ("FD&C Act"); these drugs are generally chemically synthesized, small-molecule products, not biologics.

Under Hatch-Waxman, competition from generic drugs has substantially reduced prescription drug prices and overall prescription drug expenditures, increased access to therapeutic drugs for more Americans, and hastened the pace of innovation.[3] In recent years, however, several court decisions have permitted "pay-for-delay settlements" that have reduced the procompetitive aspects of the Hatch-Waxman Act. The Commission supports legislation to prohibit these types of settlements in which the branded manufacturer pays the would-be generic entrant to abandon its patent challenge and delay entering the market with a lower cost, generic product.[4]

Hatch-Waxman does not require generic applicants to duplicate the clinical testing of drugs already proven safe and effective. Duplication of safety and efficacy information is costly, an inefficient use of scarce resources, and, as the FDA has explained, raises ethical concerns associated with unnecessary human testing.

To be approved under Hatch-Waxman, the applicant must show that its generic drug product is "bioequivalent" to (basically, the same as) the branded drug product. A bioequivalence showing is much less expensive than the clinical testing required for a branded drug product. Because the generic drug is "bioequivalent" to the branded drug, it can be safely substituted for the branded drug and expected to be as effective as the branded drug. To take advantage of generic competition, states have laws that allow pharmacists automatically to substitute a generic for a branded drug, unless a doctor has indicated otherwise.

The scientific differences between biologic and small-molecule drug products, however, complicate efforts to devise an approval process for FOB drugs based on bioequivalence. Biologic products are more complex and immunogenic than small-molecule drugs.[5] Current technology does not yet allow for the creation of an exact replica of a pioneer biologic drug product, according to the FDA. In addition, technology is not yet robust enough to determine whether an FOB product is "interchangeable" with the pioneer product such that a patient would be able to switch between the two products without the risk of an adverse effect. In light of these complexities, current legislative proposals permit FDA approval of an FOB drug that is sufficiently *similar* to, but not an exact replica of, the pioneer biologic product.[6] A showing of similarity is likely to save FOB manufacturers some clinical testing expenses but would require substantially more expense than a showing of bioequivalence for small-molecule generic drugs.

Whether competition between a pioneer biologic and an FOB is likely to be similar to competition between a branded and a generic drug is crucial to determining whether legislation to foster FOB competition should follow the same model as the Hatch-Waxman Act. Basic questions include whether the same issues that prompted provisions of the Hatch-Waxman Act that restrict entry by generic competitors are likely to be present in the context of FOB competition. To answer these questions, the Commission studied how competition between pioneer biologics and FOBs is likely to develop to determine whether similar entry restrictions would benefit consumers.

The Commission brings substantial expertise to examining likely models of competition and likely competitive effects from particular regulatory schemes.[7] To assist in its study of the issues, the Commission solicited two rounds of public comments, conducted a public roundtable discussion on November 21, 2008, and accepted additional analysis and comments through May 2009. This chapter analyzes and synthesizes the Roundtable discussion, the comments received, and relevant economic literature to assess these issues. The Commission's findings and recommendations follow.

1. Competition Between a Biologic Drug and an FOB is Much More Likely to Resemble Brand-to-Brand Competition than the Dynamics of Brand-Generic Competition under Hatch-Waxman.

Pioneer manufacturers, potential FOB manufacturers, and payors were virtually unanimous in their predictions that competition from FOB drug entry is likely to resemble brand-to-brand competition, rather than brand-to-generic drug competition. Experience to date for two markets with both pioneer biologic and FOB competitors (in Europe and the U.S.) confirms that, unlike generic drug entry, FOB entry has not resulted in steep price discounting, or rapid acquisition of market share, by FOB manufacturers.[8] This finding is true for a number of reasons:

- **The substantial costs to obtain FDA approval, plus the substantial fixed costs to develop manufacturing capacity, will likely limit the number of competitors that undertake entry with FOB products.** FOB products are likely to take eight to ten years to develop, and their development will likely cost between $100 and $200 million. These amounts differ substantially from the product development costs for small-molecule generic drugs, which typically take three to five years to develop and cost between $1 and $5 million.
- **Given these high entry costs, FOB entrants are likely to be large companies with substantial resources, and it is likely that only two to three FOB entrants will seek approval to compete with a particular pioneer biologic drug.** Current pioneer biologic drug manufacturers are likely to become FOB competitors in those markets in which they do not currently compete. Moreover, high entry costs are likely to limit FOB drug entry to markets with sales in excess of $250 million per year. The small number of likely FOB entrants contrasts significantly with the 10 or more generic entrants seen in many markets for small-molecule drugs.
- **The lack of automatic substitution between an FOB product and a pioneer biologic drug will slow the rate at which an FOB product can acquire market share and thereby increase its revenues.** In small-molecule drug markets, automatic substitution erodes a branded manufacturers' market share quickly once the first generic product enters the market. This situation is unlikely to occur in FOB markets. Unlike small-molecule generic drugs, FOB products will not be designated as "therapeutically equivalent" with the pioneer biologic drug product. The lack of therapeutic equivalence means that, like pioneer manufacturers, FOB manufacturers will have to market their products and negotiate individual contracts with purchasers.
- **An FOB drug also may have difficulty gaining market share due to concerns about safety and efficacy differences between a pioneer biologic drug and the competing FOB.** Physicians and their patients who have been taking a pioneer biologic drug may be reluctant to switch to an FOB due to a risk that the patients will react differently to the FOB than to the pioneer drug. Concerns such as these may limit FOB market opportunities to newly diagnosed patients.

- **The specialty pharmaceutical characteristics of FOBs also are likely to constrain the ability of an FOB entrant to obtain market share.** Specialty drugs, including biologic drugs, are commonly used to treat patients with severe, chronic diseases and sometimes fatal conditions. These drugs, which are primarily injected or infused, are combined with ancillary medical services and products that require specialty training for proper handling and administration. Because most biologic products are delivered to patients in clinics, hospitals, doctor's offices, or other medically supervised settings, shifting to another biologic product is typically more costly because it requires restocking of inventory and retraining of nurses and healthcare providers.
- **Biologic drugs currently are not reimbursed pursuant to strategies that payors often use to incentivize the use of lower-priced drugs; this, too, may limit market share acquisition by FOBs.** Biologic drug products are typically delivered to patients by healthcare providers as part of medical treatments (*e.g.*, dialysis treatments or oncology treatments) and reimbursed by health insurers as part of patients' medical benefits rather than pharmacy benefits. Consequently, traditional payor strategies to incentivize utilization of lower-priced drugs, including the use of co-pays and tiered formularies, are unlikely to apply to drive up the market share of FOBs. FOB pricing and market shares also are likely to be affected by the reimbursement methodologies used by Centers for Medicare and Medicaid Services ("CMS") for infused and injected drugs, which may not effectively drive share to lower-priced drugs.
- **As a result of these factors, FOB competition against a pioneer biologic drug is likely to develop as follows:** FOB entry is likely in biologic drug markets of greater than $250 million. Only two or three FOB manufacturers are likely to attempt entry for a given pioneer drug product. These FOB entrants are unlikely to introduce their FOB products at price discounts any larger than between 10 and 30 percent of the pioneer products' price. Although not as steep a discount as small-molecule generic drugs, a 10 to 30 percent discount on a $48,000 drug product represents substantial consumer savings. Pioneer manufacturers are expected to respond and offer competitive discounts to maintain market share. This price competition is likely to lead to an expanded market and greater consumer access. Nonetheless, the lack of automatic substitution will slow significant market share acquisition by FOB products. As a result, pioneer manufacturers are likely to retain 70 to 90 percent of their market share and, therefore, will likely continue to reap substantial profits years after entry by FOB drugs.

2. Existing Incentives that Support Brand-to-Brand Competition Among Biologic Drugs – Patent Protection and Market-Based Pricing – Are Likely to be Sufficient to Support FOB Competition and Biologic Innovation.

A legislative process for an abbreviated FDA approval of an FOB is likely to be an efficient way to bring FOBs to market because of the time and cost savings it provides. Given that FOB competition with a pioneer biologic drug is likely to resemble brand-to-brand competition among biologics, the question arises whether provisions that delay FOB entry and restrict competition are necessary to benefit consumers. No economic arguments suggest that such provisions are necessary to foster pioneer drug innovation or entry of interchangeable FOBs.

Brand-to-brand competition among biologics has developed without any special legislative incentives, but rather through reliance on the patent system and market-based pricing. Patent protection enables biotechnology firms to increase their expected profits from investments in R&D, thus fostering innovation that would not occur without patents' exclusionary rights.[9] Market-based pricing allows biologic drug firms to charge prices that reflect the value of the drugs to consumers and thus assists firms not only in recouping their substantial investments in biologic drugs, but also in receiving accurate market signals about the value of developing particular biologic drugs.

Market experience shows that pioneer pharmaceutical and biologic products already compete against other branded pharmaceutical and biologic entrants, and this competition benefits consumers. Currently, pioneer or first-in-class branded products engage in a race with other branded competitors to bring products to market.[10] It is likely that FOB competition similarly will develop without any special legislative incentives.

Indeed, any decision to adopt special legislative incentives that restrict competition may harm consumers. The Commission is mindful that the benefits of suppressing rivalry by either pioneer or FOB manufacturers are realized by a comparatively small number of firms who fully understand the importance of restricting competition. By contrast, the costs of restricting competition tend to be spread broadly across a large number of consumers, each of whom suffers a comparatively modest penalty compared to the relatively substantial gain realized by incumbent producers.[11] The phenomenon of highly focused benefits and broadly distributed costs gives firms a greater incentive to organize political resources to restrict competition.

a. A Twelve- to Fourteen-Year Exclusivity Period is Unnecessary to Promote Innovation by Pioneer Biologic Drug Manufacturers.

As explained earlier, pioneer biologic drug manufacturers are very likely to continue to earn substantial revenues even after the entry of FOBs. FOBs are unlikely to introduce their products at price discounts beyond 10 to 30 percent. Moreover, FOBs are likely to have difficulty rapidly growing their market shares as compared to generic small-molecule drug products. Indeed, projections are that branded biologic drugs are likely to maintain their first-mover advantages by retaining 70 to 90 percent of their market share years after FOB entry.

In addition, there is very little data to suggest that biologic drugs under development are likely to be unpatentable. Pioneer biologic drugs are covered by more and varied patents, including manufacturing and technology platform patents, than small-molecule branded products. Moreover, there is no evidence that patents claiming a biologic drug product have been designed around more frequently than those claiming small-molecule products.

Pioneer biologic manufacturers nevertheless have suggested that Congress institute a period of 12 to14 years of branded exclusivity that would begin once a pioneer biologic was approved by the FDA.[12] During this period, the FDA would be prohibited from approving an FOB product that would compete with the pioneer biologic drug. This branded exclusivity would be in addition to, and would run concurrent with, a biologic drug's existing patent protection. The economic model put forth by pioneer drug manufacturers to justify this period is based on the average time required to recoup the investment to develop and commercialize a typical biologic drug (referred to as the "Nature model").[13]

Congress has implemented exclusivity provisions in the past to encourage the development of new and innovative drug products when the drug molecule is in the public domain, and therefore not patentable. The Hatch-Waxman Act provides a five-year exclusivity period to incentivize the development of new chemical entities and it provides a three-year exclusivity period for new clinical investigations of small-molecule drugs. In other instances, Congress has implemented an exclusivity period when market-based pricing has not provided sufficient incentive to develop drug products for children or small patient populations.

Central to each of these exclusivities is a public policy trade-off: a restriction on competition is provided in return for the development of a *new* drug product or *new* use of an existing product. A 12- to 14-year exclusivity period departs sharply from this basic trade-off, because it does not spur the creation of a *new* biologic drug or indication. The drug has already been incentivized through patent protection and market-based pricing.

The potential harm posed by such a period is that firms will direct scarce R&D dollars toward developing low-risk clinical and safety data for drug products with proven mechanisms of action rather than toward new inventions to address unmet medical needs. Thus, a new 12- to 14-year exclusivity period imperils the efficiency benefits of a FOB approval process in the first place, and it risks over-investment in well-tilled areas.

The Nature model as currently structured contains numerous methodological and conceptual weaknesses that render its results too imprecise and non-robust to inform discussions about the ideal length of any branded exclusivity period. A model that balances the benefits of FOB competition (*i.e.*, lower prices and an increased pace and scope of innovation) with the costs of potentially forsaking marginal branded drug development projects would be more informative than the Nature model's approach.

Moreover, to the extent that there are new biologic molecules that cannot obtain patent protection, an exclusivity period may be warranted. Because there is no evidence about the lack of patentability of new biologic products, nor that market forces have been insufficient to incentivize their development, the Commission has not recommended a specific length for an exclusivity period.

b. Special Procedures to Resolve Patent Issues Between Pioneer and FOB Drug Manufacturers Prior to FDA Approval Are Unnecessary and They Could Undermine Patent Incentives and Harm Consumers.

Once a pioneer biologic drug manufacturer receives FDA approval and is about to market its product, it faces the risk of patent infringement litigation. FOB manufacturers are likely to face the same risk. If they believe the patent situation justifies their decisions to launch prior to resolution of any patent infringement litigation, they will enter once they have received FDA approval. If not, they will wait for the patents to expire and then launch their product. Special procedures, providing an early start to resolving patent disputes between pioneer and FOB manufacturers prior to FDA FOB approval, are not necessary to encourage FOB entry that otherwise would not have occurred.

Hatch-Waxman's special procedures for small-molecule drugs provide for an early start of patent litigation. Hatch-Waxman procedures have been the subject of extensive litigation, unintended consequences, and delayed generic entry. These procedures were designed in 1984 to address the issue of "judgment proof" generic defendants. In small-molecule drug competition, the profits of the alleged infringer (the generic entrant) are substantially less than the loss of profits by the branded product manufacturer, because of the substantial price differences between branded and generic products. Consequently, especially at the beginning of the generic industry in 1984, concerns existed that generic entrants in small-molecule drug markets might be unable to satisfy a potential treble damage award for infringing the branded manufacturer's patents.

FOB entrants will not be similarly judgment proof. FOB drug manufacturers are likely to be many of the same companies that have pioneered biologic drugs; thus, they will have the expertise and resources necessary to assess whether to launch their product before any patent infringement litigation is resolved, just as they do with a launch of a pioneer branded drug. Moreover, FOB manufacturers are highly unlikely to offer steep discounts that could jeopardize their ability to pay patent damages.

Special procedures are unlikely to be successful in providing patent certainty to the parties, because pioneer biologic drugs are covered by more and varied patents than small-molecule drugs. A special pre-approval patent resolution process is unlikely to succeed in raising and resolving all pertinent patent issues prior to FDA approval. Patents claiming the pioneer product may issue after a pre-approval process has begun and/or after FDA approval. The FOB manufacturer's application and product also may change during the approval process such that starting patent litigation prior to FDA approval would not ensure earlier resolution. Moreover, without a mechanism to enforce the rules of a pre-approval resolution process, there is no guarantee that litigation started prior to FDA approval will end earlier. In essence, early start does not guarantee early resolution.

Special procedures also could undermine the innovation incentives that patent protection affords pioneer biologic manufacturers. Although special procedures govern patent litigation between branded and generic competitors over small-molecule drug products, these procedures are the exception, not the norm.

Finally, based on the experience under Hatch-Waxman, a pre-approval patent resolution process also is likely to lead to consumer harm, including the facilitation of anticompetitive conduct that defeats the purpose of starting the patent litigation early. In the Hatch-Waxman context, branded manufacturers have used the pre-approval patent regulations to delay generic entry. In addition, generic and branded competitors have entered into "pay-for-delay" patent settlements that delay entry, not encourage it. It is likely that a pre-approval patent resolution process in the FOB context could facilitate collusive agreements and/or provide the pioneer biologic drug manufacturer with competitively sensitive information about a significant potential competitor to which it otherwise would not have access.

c. FOB Drug Manufacturers Are Unlikely to Need Additional Incentives to Develop Interchangeable FOB Products.

The question arises whether an FOB manufacturer needs an incentive beyond market-based pricing to develop an interchangeable FOB drug, such as a limit on when subsequent interchangeable FOB drug entry can occur. This limitation would allow the first interchangeable FOB manufacturer to recoup its development expenses. Because the market dynamics of FOB entry are likely to resemble competition among branded biologic drugs, provisions modeled after the Hatch-Waxman Act's 180-day marketing exclusivity are unlikely to be necessary and, indeed, could harm consumers.

The Hatch-Waxman Act provides a 180-day marketing exclusivity period to the first generic drug applicant that seeks FDA approval prior to the expiration of patents relating to the branded drug product. No other generic manufacturer may obtain FDA approval to market its product until the first generic has sold its product for 180 days or has forfeited its exclusivity period.

The 180-day exclusivity period incentivizes generic manufacturers to challenge the patents claiming a pioneer small-molecule drug product. A court finding of patent invalidity benefits not only the challenger, but also subsequent generic applicants whose entry is no longer blocked by the patent. Thus, the 180-day marketing exclusivity period prevents immediate free-riding by subsequent generic applicants on a favorable outcome that results from a generic applicant's patent challenge. As subsequent generic firms enter, generic prices can drop to 80 percent off the branded price, depending upon the number of entrants.[14] The exclusivity period is supposed to permit the first generic entrant to recoup its patent litigation costs before the substantial price drop caused by multiple generic entrants.

The competitive dynamics that justified the 180-day exclusivity period for small-molecule generic drugs are unlikely to be present here, because the entry of a subsequent interchangeable FOB is unlikely to cause a substantial price drop due to the high costs of developing and manufacturing and FOB. The first interchangeable FOB to enter will continue to earn sufficient profits even after entry of subsequent interchangeable products. Thus, market opportunities are likely to be sufficient to incentive development of interchangeable FOBs.

Not only do market dynamics counsel against an FOB exclusivity period, but the anticompetitive delay in entry evidenced in small-molecule generic drug markets is likely to repeat if an exclusivity provision for interchangeable FOBs is implemented.[15] The current 180-day exclusivity period exacerbates the problem of "pay-for-delay" settlement that prevents generic entry.[16]

Awarding an FOB exclusivity period on a "first-to-approve" rather than a "first-to-file" basis does not lessen the potential harm. These anticompetitive consequences are likely to result if the period can be extended, the period does not run immediately upon its award, or if a firm has the ability to delay triggering the running of the period through, for example, a patent settlement, acquisition, merger, or agreement.

Introduction

The Commission initiated this inquiry because decisions of regulatory bodies such as the Food and Drug Administration substantially shape business rivalry.[17] This inquiry is very mindful of how innovation in the biotechnology industry is highly dependent on patent protection.[18]

Biotechnology innovation is costly and unpredictable, requiring significant amounts of investment to test and commercialize new drug products. By preventing rival firms from free riding on discoveries, patents allow firms to recoup the substantial capital investments made to discover, test, and obtain regulatory approval of new drug products. Patents also are necessary to attract the capital to fund high-risk investment in the biotechnology industry.[19] Thus, this chapter approaches this problem by examining the likely competitive effects of a new regulatory scheme in the highly risky, costly and time-consuming process of bringing new biologic drugs to the market.

Chapter 1 of this chapter examines the likely market impact of FOB entry and contrasts it to the market impact of small-molecule generic drugs. The Commission is mindful that the likely competitive effects of FOB entry are based on the available knowledge of existing external market conditions. For example, the likely competitive effects of FOB competition could change if technology breakthroughs occur, biosimilar safety issues arise, health insurance coverage expands, or payor and reimbursement strategies change, among others. In sophisticated industries such as biotechnology, external conditions can and do change and often alter expectations of profit-maximizing firms.[20] This industry, however, has shown significant ability to adapt and thrive under new market conditions.[21] The Commission expects the robust and dynamic market conditions of the biologic drug industry to continue with the entry of FOB drug products.

Chapter 2 examines whether in addition to patent protection and market-based pricing, pioneer biologic drug products need a branded exclusivity period to promote innovation in biologic drug markets. Chapter 3 examines whether special procedures are necessary to resolve potential patent disputes between pioneer and FOB manufacturers prior to FDA approval of an FOB drug product. Chapter 4 examines whether market profits are insufficient to incentivize the development of interchangeable FOB products.

The FTC appreciates the 29 comment filers and 30 panelists who contributed time, effort, and thoughtful analysis to these issues before, during, and after the public roundtable discussion. We also are grateful for the intellectual property and economic experts proffered by the biotechnology and pharmaceutical manufacturers.

1. Background and Likely Market Impact of Follow-On Biologic Competition

I. Background

Innovations in biotechnology have improved medical treatments, reduced suffering, and saved the lives of millions of Americans. The lure of patent protection, coupled with the ability to price at market rates, has spurred pioneer drug manufacturers to develop new therapeutic drugs known as biologics.[22] The Food and Drug Administration ("FDA") approves biologic drugs under the Public Health Safety Act ("PHS Act").

These innovations, however, are expensive. As examples, annual treatment for breast cancer with the biologic drug Herceptin can cost $48,000 and the annual treatment for rheumatoid arthritis with Remicade can cost approximately $20,000. Indeed, in 2007, Americans spent $286.5 billion for prescription drugs, $40.3 billion of which was for biologic drugs.[23]

In 1984, Congress enacted the Hatch-Waxman Act to allow the FDA to approve the sale of generic or follow-on versions of off-patent branded drugs.[24] This process applies to drugs regulated only under the Federal Food Drug and Cosmetic Act ("FD&C Act"), which are generally chemically-synthesized, small-molecule products. It does not apply to drugs approved under the PHS Act.

Under Hatch-Waxman, generic applicants are not required to duplicate the clinical testing of drugs already proven safe and effective. Rather, to be approved, the applicant must show that its generic drug product is the same as the branded drug product. A bioequivalence showing is much less expensive than the clinical testing required for a pioneer branded drug product and thus, is an efficient way to leverage scarce research and development ("R&D") funds to target innovative drug development. Competition provided by the generic drug industry has reduced prescription drug prices, increased access for more Americans, and hastened the pace of innovation.[25]

There is no similar approval process for biologic drugs.[26] Rather, once a biologic drug product's patents expire, the follow-on applicant must duplicate the clinical testing of the pioneer biologic drug. This duplication of safety and efficacy information is costly, an inefficient use of scarce resources, and, as the FDA has explained, raises ethical concerns associated with unnecessary human testing.

The desire to avoid these consequences by creating an approval process for follow-on biologic ("FOB") drugs takes on urgency in light of the significant number of biologic drugs that go off-patent within the next several years. Figure 1-1 shows the 27 top selling biologic drug products, many of which go off patent by 2015.[27] The drugs listed comprise approximately 87 percent of the total global value of the biologics industry of $112 billion.

The scientific differences between biologic and small-molecule drug products, however, complicate efforts to devise an approval process for FOB drugs. Biologic products are more complex and immunogenic than small-molecule drugs.[28] Current technology does not yet allow for the creation of an exact replica of a pioneer biologic drug product, according to the FDA.[29] In addition, technology is not yet robust enough to determine whether an FOB product is "interchangeable" with the pioneer product such that a patient would be able to switch between the two products without an adverse effect.

In light of these complexities, current legislative proposals permit FDA approval of an FOB drug that is sufficiently *similar* to, but not an exact replica of, the referenced branded biologic product.[30] A showing of similarity is likely to save clinical testing expenses but would require substantially more expense than a showing of bioequivalence for small-molecule generic drugs. Unlike small-molecule drugs, FOB products would not be designated as "therapeutically equivalent" with the referenced product. The lack of therapeutic equivalence means that a pharmacist may not substitute prescriptions for a pioneer product to an FOB product without physician consent. As technology and scientific understanding develops, however, the approval process could provide a means by which an FOB applicant could show that its product is interchangeable with the pioneer product.

In the current legislative debate, questions have arise over whether the same issues that prompted provisions of the Hatch-Waxman Act that restrict entry by generic competitors are likely to be present in the context of FOB competition. To answer these questions, the Commission initiated a public inquiry, including a public workshop and a series of public comments, to examine how FOB competition is likely to develop to determine whether similar entry restrictions would benefit consumers.[31]

This chapter describes the regulatory background necessary to understand how an FOB approval process could be used by FOB manufacturers. It then describes the likely market impact of FOB entry and contrasts it to the market impact of small-molecule generic drugs. This analysis sets the stage for the discussion in Chapters 2 through 4 of specific issues regarding how to foster FOB competition to benefit consumers.

Drug	2008 Sales	Year Approved	Drug	2008 Sales	Year Approved	Drug	2008 Sales	Year Approved
Avastin	$9.2	2004	Novolog	$3.7	2000	Rebif	$1.7	2002
Enbrel	$8.0	1998	Erbitux	$3.6	2004	Cerezyme	$1.5	1994
Remicade	$7.9	1998	Aranesp	$3.2	2001	Tysabri	$1.4	2004
Humira	$7.3	2002	Recombinate	$2.9	1998	NovoSeven	$1.4	1999
Rituxan	$7.3	1997	Lucentis	$2.7	2006	Synagis	$1.3	1998
Herceptin	$5.7	1998	Avonex	$2.6	1996	Neupogen	$1.3	1991
Lantus	$5.1	2000	Novolin	$2.5	1991	Betaseron	$1.2	1993
Epogen/ Procrit	$5.1	1989	Humalog	$2.2	1996	Humulin	$1.1	1992
Neulasta	$4.2	2002	PEGASYS	$2.0	2002	Kogenate FS	$1.1	1993

Figure 1-1. Top-Selling Biologic Products (estimates of peak sales in billions)

II. The New Drug and Generic Approval Processes

A. New Drug Approval Processes Under the FD&C Act and the PHS Act

To obtain FDA approval of a new small-molecule drug under the FD&C Act or a biologic product under the PHS Act, the manufacturer must prove that the product is safe and effective. Manufacturers must submit the following information to the FDA for approval:

(a) pre-clinical analytical tests, pre-clinical studies and formulation studies;
(b) an Investigational New Drug Application ("IND") to initiate human clinical testing;
(c) adequate and well-controlled human clinical trials to establish the safety and efficacy of the drug for its intended use;
(d) approval and validation of commercial scale manufacturing facilities used in production of the product;
(e) drug manufacture and analytical methods; and
(f) proposed product packaging and labeling.[32]

The pre-clinical phase of any new drug development typically identifies compounds (either small-molecule or protein-based) that target a particular disease or are therapeutically beneficial. Once a lead compound is isolated, the manufacturer conducts pre-clinical safety trials, as well as trials in predictive animal models to determine if the compound works as expected. This pre-clinical phase typically takes one to five years.[33]

After pre-clinical tests are completed, a drug sponsor submits these results in an IND to the FDA before human clinical trials may commence.[34]

Clinical trials typically consist of three phases. In Phase I, a small group of patients is given the drug to determine if the drug is safe in humans. In Phase II, a small sample of the intended patient population is given doses of the drug to provide a preliminary assessment of the efficacy of the drug for a specific clinical indication, find dose tolerance, and find the optimal dose range. Phase III studies are initiated if Phase I and Phase II studies indicate the drug is safe and has some efficacy in the targeted patient population. Phase III studies are designed to gather sufficient data in a broad target population in order to establish safety and efficacy for a particular indication.

The time to conduct these trials varies based on factors such as indication, availability of reliable ways to measure efficacy, size of patient populations in the clinical trials, ease of patient accrual, as well as a host of other factors. Despite these variances, Phase I takes approximately one year, Phase II (including dose ranging studies) takes approximately two years, and Phase III takes approximately three years.[35]

B. Generic Drug Approval Under the FD&C Act

Rather than requiring a generic manufacturer to repeat the costly and time-consuming new drug approval process, the Hatch-Waxman Act permits generic drug applicants to file an Abbreviated New Drug Application ("ANDA"). The object of the ANDA process is to demonstrate that the generic drug product has the same active ingredient, route of administration, dosage form, strength, and proposed labeling as the branded drug. The ANDA also must contain sufficient information to demonstrate that the generic drug is "bioequivalent" to the relevant branded product.[36] As a result of providing this information, the generic applicant may rely on the FDA's previous findings of safety and effectiveness for the branded drug, and the applicant, therefore, does not have to perform its own clinical studies. This reliance allows generic applicants to save substantial time and development costs.[37] The FDA will deem a generic drug product therapeutically equivalent to the branded product. This designation allows the generic drug to be automatically substituted by a pharmacist for the branded product.

C. Issues in Translating the Generic Drug Approval Process to Biologic Drugs Under the PHS Act

The scientific differences between biologic and small-molecule drugs complicate efforts to devise an approval process for FOB drugs based on bioequivalence.[38] Figure 1-2 shows the size differences between a typical small-molecule drug and a biologic protein and lists some of the complexities surrounding protein drugs. These differences include a ten to hundred-fold difference in size. A small-molecule drug, such as a statin (*e.g.,* Lipitor, Mevacor), is small (only 400 Daltons) and simple in contrast to a biologic drug. A biologic drug is significantly larger (5,000-300,000 Daltons) and has a complex structure with three-dimensional folding which performs complex binding, unlike small-molecules. Any deviation in a biologic protein's structure can result in aggregation, incorrect folding and structural anomalies (*e.g.*, truncation, proteolysis and amino acid modifications) that can have unexpected effects on efficacy and safety.[39]

Current limitations in analytical methods make it difficult to characterize and compare large molecules to determine their level of sameness. Manufacturing a consistent biologic drug product presents additional difficulties.[40] In light of these challenges, it is unlikely that FOB manufacturers could only use analytic methods to show that their FOB products have the same active ingredient as the pioneer biologic product, as generic small-molecule drug applicants do pursuant to the Hatch-Waxman Act.[41]

In light of these complexities, current legislative proposals permit FDA approval of an FOB drug that is sufficiently *similar to*, but not an exact replica of, the pioneer product.[42] A showing of similarity is likely to save clinical testing expenses but would require substantially more expense than a showing of bioequivalence for small-molecule generic drugs. The amount of savings, however, may vary depending upon the complexity of the pioneer product to ensure that the FOB product is safe, pure and potent.[43] Although abbreviated compared to a full development program, FOB applicants are likely to perform Phase I and Phase III studies, but with fewer patients. FOB manufacturers also must seek approval and validation of their commercial-scale manufacturing facilities at or before initiation of clinical trials.[44] For each additional indication for which they seek labeling, FOB manufacturers are likely to be required to perform Phase I – Phase III clinical testing.[45]

Unlike small-molecule drugs, FOB products would not be designated as "therapeutically equivalent" with the referenced product. The lack of therapeutic equivalence means that a pharmacist may not substitute prescriptions for a pioneer product to an FOB product without physician consent. The approval process could provide, however, a means by which an FOB applicant could show that its product is interchangeable with the pioneer product as technology and scientific understanding develops.[46]

It also is likely that FOB manufacturers could become innovators. For example, they may develop "biobetter" FOB drugs that improve upon the safety and effectiveness of the pioneer product. In other instances, FOB firms could develop improved manufacturing processes and analytics, resulting in safer biologics manufactured by both pioneer and FOB manufacturers, and/or more efficient manufacturing and testing methodologies, resulting in lower-priced biologic drugs.[47] One commenter suggested that the "incentive for enhanced and innovative biologics manufacturing capacity is an oft-forgotten but critically-important aspect of innovation, particularly in the context of biologics, and it is one that can enable a direct reduction in the cost of goods and an increased durability of supply."[48]

III. Pharmaceutical Pricing, Market Dynamics and the Likely Competitive Impact of Follow-On Biologics

Pioneer manufacturers, potential FOB manufacturers, and payors explained that it is likely that an FOB approval process under the PHS Act will result in the approval of biosimilar products, not interchangeable ones. This section describes the likely market effects of biosimilar product entry and contrasts it to the market effects of entry by small-molecule generic drugs.

A. Pharmaceutical Pricing and the Effect of Generic Drug Entry

In the United States, a pioneer manufacturer of either small-molecule or biologic drugs is free to charge a monopoly price for its product to the extent the market conditions permit or it is perceived to offer greater health benefits compared to existing drugs or medical treatments.[49] Patent-protected drug products also may be able to prevent the manufacturer from facing competition, thus enabling the manufacturer to charge a monopoly price.

Manufacturers of small-molecule and biologic drugs market their products through a variety of channels including a specialty detail sales force, free samples or prescription coupons, medical education and conferences, peer review journal publications, direct-to-consumer advertising, and formulary access. Formulary access is controlled either by private prescription benefit managers ("PBMs") for reimbursement by health insurance companies or managers for coverage by various public payors (*e.g.*, Department of Veterans Affairs, Medicare, state Medicaid programs).[50]

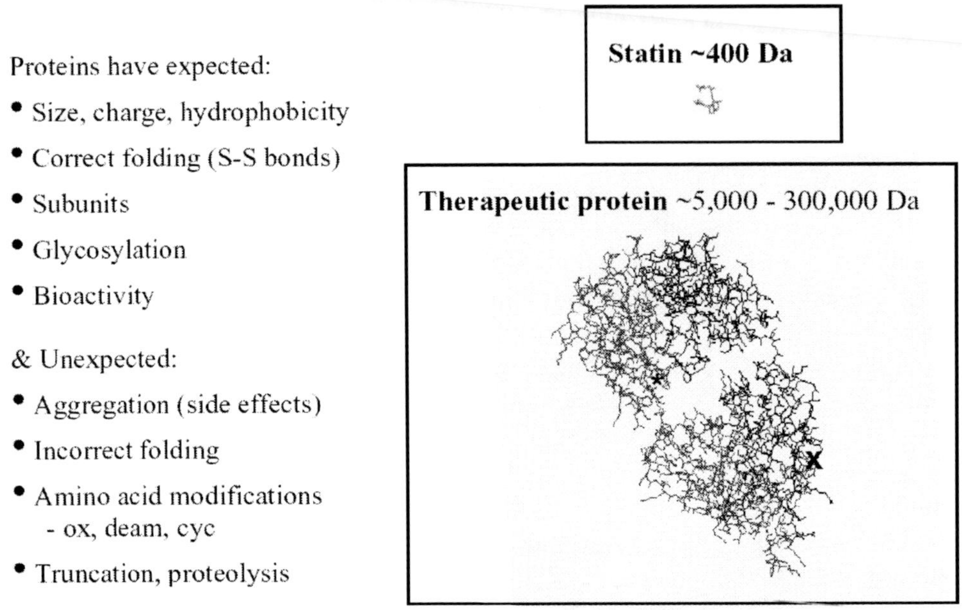

Figure 1-2. Structure of Small-Molecule *vs.* Protein Drugs

Approval of a breakthrough or pioneer drug product is increasingly followed by entry of a subsequent branded product(s).[51] The head start that the breakthrough product has had over subsequent branded products has decreased over the past three decades from 8.2 years during the 1970s to 2.25 years in the 1990s.[52]

When the FDA approves a branded competitor, price competition ensues, market size expands, and market share shifts among the competitors.[53] Brand-to-brand competition results in negotiated price discounts in the range of 18 to 27 percent off the pioneer's product price.[54] Brand-to-brand competition also expands the market (in units and dollars) for a therapeutic class of drugs by increasing awareness of conditions and treatments from increased detailing, advertising, and marketing, as firms compete to influence physician prescribing behavior in favor of their brands.[55] Price competition among branded firms therefore increases access for patients.

For drugs approved under the FD&C Act, generic entry occurs when patent protection ends[56] (either at patent expiration or by a court finding of non-infringement or invalidity). The number of generic entrants after patent expiration is largely a function of fixed entry costs compared to the market opportunity.[57] The first generic entrant generally offers a price that is 25 percent lower than the branded drug's price. The price discount can rise to 80 percent with multiple generic entrants.[58]

Marketplace experiences have documented the rapid erosion of a branded drug's sales once the first generic product is introduced.[59] The rapid decline of the branded product's market share is largely a function of state substitution laws and price sensitive customers' use of drug formularies.[60] State substitution laws allow a pharmacist to dispense a generic drug when presented with a prescription for its branded equivalent, unless the physician or consumer directs otherwise. In addition, PBMs and retail pharmacies have substantial incentive to dispense generic drugs because the margins on generic drugs are greater than they are for branded products, resulting in greater profits for PBMs and retail pharmacies.[61]

These two factors enable the generic entrant to erode a majority of the market share of the branded product within the first year.[62] When additional generic firms enter, they compete against incumbent generic firms for market share, not the branded manufacturer, because the first generic firm has already obtained most of the branded manufacturer's sales.

B. Likely Market Effects of Biosimilar Entry

Competition from FOB drug entry is likely to resemble brand-to-brand competition rather than generic drug competition.[63] Experience to date for two products with both branded and FOB competitors (in Europe and the U.S.) shows that four factors have dampened substantial price discounting by, and rapid share shifting to, FOB manufacturers as compared to the effects of generic drug entry. As a result, branded manufacturers are likely to continue to reap profits after FOB entry.

1. Fewer FOB Competitors Due to High Barriers to Entry

Fewer FOB competitors are expected due to the technological barriers and the high cost of entry.[64] FOB products are likely to take eight to 10 years to develop and to cost between $100 and $200 million.[65] Higher development costs for FOB products, compared to small-molecule generic drugs, include those associated with manufacturing, clinical trials, and post-marketing surveillance.[66] By contrast, small-molecule generic drugs product development costs range from approximately $1 to $5 million.

Follow-on biologic manufacturers will likely have to build, equip and qualify their own manufacturing facilities, which is likely to cost $250 to $1 billion.[67]

Additionally, biologic manufacturing is costly, difficult and often requires acquiring or duplicating proprietary cell lines that are protected by both patents and trade secrets. These barriers further reduce the number of likely successful FOB entrants.

In addition to the development and manufacturing costs, FOB competitors are likely to engage in marketing and sales support for their FOB products.[68] These high costs are likely to limit FOB drug entry to markets with sales in excess of $250 million per year.[69]

In light of these high entry costs, FOB entrants are likely to be large companies with substantial resources. Current biologic drug manufacturers are likely to become FOB competitors in those markets in which they do not currently compete. Potential FOB entrants could include well-established biotechnology, and hybrid biopharmaceutical firms such as: Abbott, AstraZeneca (acquisition of MedImmune and CAT), Baxter (acquisition of Knoll), Biogen/IDEC, Eli Lilly (acquisition of Imclone), Johnson & Johnson (acquisition of Centocor), Pfizer (recent announced acquisition agreement with Wyeth), Roche (acquisition of a majority interest in Genentech), Novo Nordisk, and Sanofi-Aventis.[70]

FOB firms in Europe who have an interest in developing FOB products for the U.S. market include: Novartis (including its generics division Sandoz), Teva, Hospira (partnering with German generics firm Stada), and Momenta (partnering with Novartis). Additionally, commenters recognized branded pharmaceutical firms such as Merck, Boehringer Ingelheim, and Wyeth.[71]

2. Lack of Interchangeability

The lack of interchangeability and automatic FOB substitution are likely to dampen how quickly an FOB manufacturer acquires market share compared to generic drug entry.[72] In small-molecule drug markets, automatic substitution erodes a branded manufacturers' market share quickly once the first generic product enters the market. As more generic products enter, they compete for market share among themselves, as the branded manufacturer already has lost its market share to the first generic entrant. This situation is unlikely to occur in FOB markets as FOB manufacturers will be required to market their products and negotiate individual contracts with purchasers in competition with the branded manufacturer's product.[73] FOB market share is likely to depend on: order of entry into the market; clinical trial results; size of detailing sales force; direct-to-consumer advertising; and access to formularies, which include price discounts to the most sophisticated, price-sensitive customers.[74]

FOB market penetration also is likely to be hampered by lingering or institutionalized uncertainty about interchangeability and safety differences between pioneer and FOB products.[75] This uncertainty may be heightened if the FOB product does not share the same name as the pioneer biologic product.[76] Physicians and their patients who have been safely taking a pioneer biologic drug product may be reluctant to switch to an FOB product because of the risk that the patient will react differently to the new drug.[77] These concerns may limit the FOB market opportunities to newly diagnosed patients or patients who had not improved by using the pioneer biologic drug. These concerns may dissipate as providers become more experienced with FOBs.[78]

3. Specialty Pharmaceutical Characteristics

The specialty pharmaceutical characteristics of FOB drugs also are likely to constrain market share acquisition.[79] Specialty drugs, including biologic drugs, are commonly used to treat patients with severe, chronic diseases and sometimes fatal conditions. These drugs, which are primarily injected or infused, are combined with ancillary medical services and products which require specialty training for proper handling and administration.[80] Because most biologic products are delivered to patients in clinics, hospitals, and doctor's offices, or other medically-supervised settings, shifting to another biologic product is typically more costly because it requires restocking inventory and retraining nurses and healthcare providers.[81]

4. Fewer Payor Strategies to Incentivize Rapid Uptake of FOBs

Biologic drug products are typically delivered to patients by healthcare providers as part of medical treatments (*e.g.*, dialysis treatments or oncology treatments) and reimbursed by health insurers as part of patients' medical benefits rather than the pharmacy benefits. This situation contrasts with small-molecule drug products which are dispensed by pharmacists to the patients and reimbursed by the insurance providers as part of the patients' pharmacy benefit.[82]

Traditional payor strategies used to manage pharmacy benefits that incentivize rapid shifting of patients from branded drugs to lower-priced generic drugs – for example, by requiring higher co-pays from patients for drugs off the formulary – are likely to be of limited use for biologic drugs. Consequently, payors will have fewer strategies to incentivize the rapid uptake of lower-priced FOBs, especially biosimilars.[83] In addition, the reimbursement methodologies used by Centers for Medicare and Medicaid Services ("CMS") for biologic drugs are likely to be important factors affecting the market impact of FOBs and pricing of FOBs.[84]

Because of these four characteristics, payors, branded manufacturers, and FOB manufacturers forecasted that pioneer manufacturers are likely to maintain market share for several years even after FOB entry. They predicted that market share acquisition by FOBs would be modest, lagging substantially behind the sometimes blistering competitive pace established by generic small-molecule entrants.[85] Several commenters concluded that uptake of FOBs will likely be "slower and less extensive than for many small-molecule drugs."[86] They estimated that the uptake for FOBs will range between 10 percent and 30 percent.[87] They also noted that the market share uptake of FOBs will be correlated to the price which in turn is affected by the sums needed to generate clinical trial data required by the FDA to obtain approval.[88]

Panelists noted that as the market gained positive experience with FOBs, market uptake of FOBs could increase.[89] Conversely, they also predicted that if the market had negative experiences with FOBs from safety or efficacy issues (immunogenicity, heparin like contamination or problems akin to the generic drug scandals of the 1980's), then FOB uptake could also be significantly dampened.[90]

C. Market Experience with Biosimilar Entry

Market experience with both pioneer and FOB competitors confirms that FOB competition is likely to resemble branded competition rather than generic competition as seen for small-molecule drug products. The European Union adopted an approval process for follow-on biologics in 2004.[91] To date, the European Medicines Agency has approved biosimilars for three products: (1) EPO (erythropoeitin stimulating agent or "ESA") to treat anemia; (2) human growth hormone ("HGH") to treat children with small stature, and other conditions associated with deficiencies of the naturally occurring hormone; and (3) G-CSF (Granulocyte-Colony Stimulating Factor) to stimulate production of white blood cells needed to fight infection. In the U.S., the FDA has approved two biosimilar HGH products pursuant to the FD&C Act.[92] The following two sections describe the market competition for EPO and HGH.[93]

1. EPO Market Experience in the European Union

Panelists and commenters explained that seven EPO biologic manufacturers market their products in Germany, three of which are biosimilars (products 5-7):

(1) Amgen's Aranesp,
(2) Johnson & Johnson's Eprex/Erypo,
(3) Roche's NeoRecormon,
(4) Roche's Mircera,
(5) Hospira's Retacrit,
(6) Novartis' Binocrit, and
(7) Shire's Dynepo.[94]

As of November 2008, the multiple biosimilar entrants had attained a combined market share in Germany of between 14 to 30 percent with price discounts estimated at about 25 percent off the branded price several years after entry.[95] The reported results of international sales from the first quarter of 2009 appear to confirm that pioneer firms retain a significant first mover advantage. For example, Amgen states that Aranesp's market share in dialysis patients has increased slightly even two years after biosimilar and other branded competitors have entered the market.[96]

2. HGH Market Experience in the European Union and the United States

Panelists and commenters also discussed the limited price competition and market share shift to biosimilars in the HGH markets in the E.U. and U.S. In April 2006, Novartis launched its biosimilar HGH product, Omnitrope, which referenced Pfizer's Genotropin, in Germany and Austria. In December 2006, BioPartner launched the second HGH product, Valtropin, in the E.U., which referenced Eli Lilly's Humatrope.

By leveraging its global R&D, Novartis launched Omnitrope in the United States in 2007.[97] The second HGH biosimilar entrant in the United States was Teva with Tev-Tropin.[98] There are five other branded HGH products in the U.S. market:

- Pfizer's Genotropin
- Eli Lilly's Humatrope
- Novo Nordisk's Norditropin
- Serono's Saizen
- Genentech's Nutropin (Genentech, majority-owned by Roche)[99]

As of November 2008, combined U.S. market shares of the two biosimilars amounted to about four percent.[100] Panelists' best estimates of the price discounts in the U.S. for HGH biosimilar drug products ranged from 10 to 40 percent off the branded HGH products' prices depending upon the purchaser, while branded HGH prices had increased.[101] As discussed above, pioneer manufacturers offer discounts to their most price sensitive, sophisticated, and largest purchasers; these discounts are negotiated individually and typically are not publicly available.[102] Both Novartis and Teva supported their biosimilar products with marketing and sales efforts.[103]

D. Likely Pricing Effect of Interchangeable FOBs

Panelists and commenters expressed a range of price discount predictions if and when technology allows interchangeable FOBs to enter the market. For ease of discussion, some panelists and commenters referred to interchangeable FOBs as "biogeneric" drugs. Panelists predicted that if biogeneric applicants could, for example, rely on analytical data rather than clinical trials to show equivalent efficacy, and not be required to engage additional comparability and immunogenicity trials, then biogenerics will generate greater consumer savings than biosimilars.[104] And conversely, if a biogeneric pathway were more costly and rigorous than the process for new drug approvals, panelists predicted no biogeneric FOB entrants would use such a pathway as "manufacturers would be better off pursuing a full approval."[105]

One commenter explained that savings in marketing and selling expenses should translate into lower sales price for a biogeneric product than a biosimilar product.[106] An FOB manufacturer explained that only interchangeable biogeneric, not biosimilar, products offer the greatest price competition.[107] This increased price competition, however, is likely to be greater than price competition among biosimilars, but not as great as generic drug price competition seen with small-molecule generic drugs.[108]

However, at least one panelist disagreed stating, "[i]nterchangeability will not necessarily provide greater economic benefit from biosimilar market entry." He asserted that this prediction is erroneous because it is based on the false assumption that biogeneric products would be "interchangeable" and approved without more clinical testing than biosimilars.[109] One commenter stated that not only was biogeneric entry not possible, but the effects on cost savings provided by biogenerics were too speculative to predict at this point.[110]

E. Conclusions about the Likely Market Impact of FOB Entry

An abbreviated approval process for follow-on biologic drugs is likely to be an efficient way to bring a biosimilar drug product to market. The FOB applicant can save time and money by not engaging in the full pre-clinical and clinical tests and, as a result, it can enter the market at a price lower than the pioneer drug product.

Competition from FOB drug entry is likely to resemble brand-to-brand competition rather than brand-to-generic drug competition for small-molecule products. Two or three FOB manufacturers are expected to seek entry in large markets due to the significant time and expense expected to develop an FOB drug product. They are likely to introduce their drug products at price discounts between 10 and 30 percent of the pioneer products' price to the most price-sensitive customers. Pioneer manufacturers are expected to respond aggressively and offer competitive discounts. This price competition is likely to lead to an expanded market and greater consumer access.

The lack of automatic substitution will slow significant market share acquisition by FOB products. The difficult and costly administration, training, payment, and reimbursement of specialty drugs makes it likely that there will be few entrants, despite the multi-billion dollar size market opportunities offered by many biologic products losing their patent protection in the next 10 years. Moreover, traditional payor incentives used in the retail pharmacy setting, such as co-pay differential and formulary tiering to incentivize utilization of low-priced drugs, are unlikely to be used in the specialized drug setting in which many biologics are dispensed, such as hospitals and outpatient clinics, and other clinical settings. As a result, pioneer manufacturers are likely to retain 70 to 90 percent of their market share after FOB entry.

The likely effect of FOB entry contrasts markedly from small-molecule generic drug competition. Soon after small-molecule generic drug entry occurs, the branded product loses most of its market share. This loss of market share occurs because of state substitution laws and payor incentives that permit pharmacies to substitute a prescription for a branded product to a generic product without physician consent. When a market includes eight or more generic products, prices can be discounted up to 80 percent of the branded price.

The Commission is mindful that the likely competitive effects of FOB entry described in this chapter are based on agreement among pioneer manufacturers, potential FOB applicants, and payors as to future conditions. The likely competitive effects of FOB competition could change if technology breakthroughs occur, biosimilar safety issues arise, health insurance coverage expands, or payor and reimbursement strategies change, among others. In sophisticated industries such as biotechnology, external conditions can and do change and often alter expectations of profit-maximizing firms.[111] This industry, however, has shown significant ability to adapt and thrive under new market conditions.[112] The Commission expects the robust and dynamic market conditions of the biologic drug industry to continue with the entry of FOB drug products.

2. Patent Protection and Market Incentives are Likely to Continue to Provide Robust Innovation Incentives after Entry of Follow-On Biologic Drugs

I. Introduction

The introduction of FOB competition raises the question of whether, in addition to patent protection and market-based pricing, pioneer biologic drug products need an exclusivity period, a "branded exclusivity period," that restricts FOB competition by prohibiting the FDA from approving an FOB product for some period of time to promote innovation in biologic drug markets.[113] Pioneer biologic drug manufacturers have suggested that a 12- to 14-year branded exclusivity period is necessary to incentivize innovation.[114] The length of this branded exclusivity period is based on a model that estimates the time it takes a pioneer manufacturer to recoup its investment to develop and commercialize a typical biologic drug (the "Nature model").[115]

This chapter explains that the main argument for a branded exclusivity period of 12 to 14 years is to compensate for the perceived failures of the patent system to reward, protect, and incentivize biologic drug innovation.[116] To understand whether such a branded exclusivity period is necessary, and the likely effects of such a period, this chapter summarizes the comments and relevant economic literature on how biologic drugs are developed and the role of the patent system in driving these innovations.

The patent system is the primary means by which the government grants exclusive rights to promote innovation. Patent protection and market-based pricing enables biotechnology firms to increase their expected profits from investments in R&D, thus fostering innovation that would not occur without patents' exclusionary rights.[117] Congress and the courts set patent policy with a conscious eye towards maintaining an appropriate balance with competition policy, which also promotes innovation, as the best means to benefit consumers.[118]

Nothing about the introduction of FOB drug products changes the relationship of pioneer biologic drug products to the patents protecting them. As a result, patent protection should continue to incentivize biotechnology innovation, even after enactment of an approval process for FOB drugs. Pioneer biologic drugs are covered by more and varied patents than small-molecule branded products, including manufacturing and technology platform patents. Moreover, there is no evidence that patents claiming a biologic drug product have been designed around more frequently than those claiming small-molecule products.

Even if the FOB manufacturer were to design around the patents claiming a pioneer biologic drug product and enter prior to patent expiration, the effect of FOB entry is unlikely to cause the precipitous decline in the pioneer product's revenues that generic drug entry causes. FOB drugs are likely to garner only 10 to 30 percent market share of an expanded market, rather than nearly 100 percent of the market share from a branded small-molecule drug manufacturer. The pioneer biologic drug manufacturer can continue to earn significant revenues years after FOB entry.

The use of patents to incentivize innovation is especially strong if the FOB approval process does not contain special features similar to the ones in Hatch-Waxman that incentivize an early start to patent challenges that is prior to FDA approval of the generic drug. (These issues are discussed in Chapters 3 and 4.) These early patent challenges are unique to the generic drug industry and, if applied in the FOB drug context, undermine the ability of the patent to incentivize innovation.

Market experience shows that pharmaceutical products already compete against other branded entrants and that this competition benefits consumers by increasing the pace and scope of innovation as well as price competition. Currently, pioneer or first-in-class branded products engage in a race with other branded competitors to bring products to market.[119] Over the last three decades, the head start of the first-in-class drug product has decreased as the average lead time of the first-in-class product shrank from 8.2 years during the 1970s to 2.25 years in the 1990s. This limited head start for the first-in-class drug product has not dampened R&D incentives and may, in fact, be optimal for rewarding past innovation while allowing competition to incentivize future innovation.[120] Because FOB entry is likely to have a competitive effect similar to that caused by entry of another branded competitor, it is likely that FOB entry will have a similar effect on innovation.

Congress has implemented exclusivity periods to encourage the development of new and innovative drug products when the drug molecule is in the public domain, and therefore not patentable. The Hatch-Waxman Act provides a five-year exclusivity period to incentivize the development of new chemical entities and it provides a three-year exclusivity period for new clinical investigations ("NCI") of small-molecule drugs. In other instances, Congress has implemented an exclusivity period when market-based pricing has not provided sufficient incentive to test drug products for children or small patient populations.

Central to each of these exclusivities is a public policy trade-off: a restriction on competition is provided in return for a development of a new drug product or new use of an existing product. A 12- to 14-year exclusivity period for pioneer drugs, however, departs sharply from this basic trade-off, because it does not spur the creation of a new product or indication. The product has already been incentivized through patent protection and market-based pricing.

The potential harm posed by such a period is that firms will direct scarce R&D dollars toward developing low-risk clinical and safety data for drug products with proven methods of action rather than toward new inventions to address unmet medical needs. Thus, a new 12- to 14-year exclusivity period imperils the efficiency benefits of an FOB approval process in the first place and it risks over-investment in well-tilled areas.

This chapter then summarizes a critique of the Nature model. The model as currently structured contains numerous methodological and conceptual weaknesses that render its results too imprecise and non-robust to inform discussions about the ideal length of any branded exclusivity period. A model that balances the benefits of FOB competition with the costs of potentially forsaking marginal branded drug development projects would be more informative than the Nature model's approach.

II. The Importance of Patent Protection to the New Drug Research and Development Process

A. New Drug Research and Development Process

Pharmaceutical innovation for new drug products is lengthy, expensive, highly risky, and involves a multitude of public- and private-sector entities.[121] Pharmaceutical innovation begins with basic scientific research. Much of the funding of basic medical research comes from the National Institutes of Health or other government sources, angel investors and corporations, not venture capitalists.[122] This funding covers basic research up until proof of concept, which is usually demonstrated by preclinical findings.[123] This basic medical understanding of disease pathways and processes has led to the commercialization of two categories of biologic drugs: (1) recombinant proteins; and (2) monoclonal antibodies.[124]

Once proof of concept has been attained, private investment from angel investors, corporate, and venture capital funding continues the development of these inventions through clinical development of a drug candidate.[125] After late stage clinical development, private corporations typically begin to scale-up manufacturing and marketing efforts.[126] The R&D process for developing biologic drugs lasts, on average, 10 to 12 years.[127]

Pioneer biologic manufacturers also engage in a race to screen, patent, and develop their products.[128] These races are often propelled by a new medical threat or scientific advances that suggest a new line of therapy.[129]

A study of first-in-class drugs approved by the FDA from the 1960s through the 1990s shows that increasingly, multiple firms target the same disease, therapy or biologic pathway, and as a result, nearly every therapeutic class has had multiple branded competitors. Branded competitors' R&D occurs in parallel. For example, in the 1990s, for all drug classes in which a first-in-class drug was approved, clinical testing for at least one branded competitor's drug occurred before FDA approval of the first-in-class drug.[130] The head start of the first-in-class drug product has decreased over the last three decades, shrinking markedly from 8.2 years during the 1970s to 2.25 years in the 1990s.[131]

Competition does not stop once FDA approval is obtained. Biologic drug manufacturers in particular seek to expand the market opportunity for their products by obtaining additional indications for diseases that share biologic pathways; for example, HGH indications for Turner's syndrome and pituitary dwarfism, Tumor Necrosis Factor ("TNF") inhibitors for both Crohn's Disease and rheumatoid arthritis, and Vasoendothelial Growth Factor ("VEGF") inhibitor for lung cancer and colorectal cancer. These incremental innovations lead to "improvements that over time can yield substantial benefits."[132]

B. The Importance of Patent Protection Incentives for Innovation

Patent protection fuels this R&D engine.[133] To obtain a patent, an invention (*i.e.*, a product, a process, machine, or composition of matter) must be novel, non-obvious, and useful. A patentee also must disclose clearly the invention. Economic literature has described how this property right enables biotechnology firms to increase their expected profits from investments in R&D, thus fostering innovation that would not occur but for the prospect of a patent.[134]

The FTC, in its 2003 Report, "To Promote Innovation: The Proper Balance of Competition and Patent Law and Policy," described how innovation in the biotechnology and pharmaceutical industries is highly dependent on patent protection, more so than any other industry.[135] Stand-alone innovation in these two industries is costly and unpredictable, requiring significant amounts of pioneering research to discover and test new drug products. By preventing rival firms from free riding on discoveries, patents allow pioneer firms to recoup the substantial capital investments made to discover, test, and obtain regulatory approval of new drug products. Patents also are necessary to attract the capital to fund high-risk investment in the biotechnology industry.[136]

The FTC Patent Report explained how pharmaceutical and biotech firms use the patent information disclosures required by the patent statutes to direct their R&D into areas not claimed by patents. Patent disclosures can guide rival firms' efforts to "design-around" patents, so that they can develop non-infringing products to compete with the patented discovery and thus spur greater innovation.[137]

Patent protection also covers several components of biotechnology products, including claims drawn to:[138]

- the compound or molecule,
- methods of treatment (specific indications, route of administration),
- formulation and dosage form,
- product-by-process claims (products defined by the process used to make the molecule),
- manufacturing process (including cell lines used in the manufacturing process), and
- manufacturing technology (technology platforms and research tools used to make the molecule).

With one key difference, these are the same types of patent claims that claim small-molecule products. Process patents and technology platform patents are often more important for biologic drug products than for small-molecule drug products. Process patent claims are important because the "processes by which biologics are made are highly specific, complex, and determine many of the biologic's functional and structural characteristics. . . [that] can often be expected to affect the product's safety, purity, and efficacy profile, and thus are integral to the approval of the product itself."[139] Process claims, therefore, add a layer of patent protection that small-molecule drug products may not possess.[140]

Indeed, concern previously centered on the belief that biotechnology patent protection was too strong because of patent claims covering research tools used to assist in the drug discovery process. The concern was that patented research tools would actually obstruct commercialization of new products, thereby hindering follow-on innovation.[141] This problem has yet to materialize.[142]

C. Patent Protection in the Biotechnology Industry

The introduction of FOB competition raises the question of whether pioneer biologic drug products should be granted an exclusivity period to incentivize innovation. This section summarizes the two competing arguments regarding the strength of patents to continue to incentivize biotechnology R&D in the face of FOB competition.

1. Panelists' and Commenters' Arguments that Patents Are Unlikely to Incentivize Innovation in Light of FOB Competition

Panelists and commenters representing pioneer biologic drug manufacturers suggested that biologic drug patents are likely to provide less investment certainty than patents claiming small-molecule drug products because FOB products are likely to be similar to, not exact duplicates of, the branded drug product.[143] The panelists suggested that FOB competitors could develop biosimilar products by designing around the branded product's patents.[144]

A commenter predicted that the pioneer drug manufacturer does not know whether its patent estate is going to cover the exact molecule that an FOB manufacturer produces.[145] Another commenter summed up this difficulty by suggesting that "the uncertain 'similarity' standard for approval of FOBs creates a greater potential for biologic patents to be designed around, particularly given some of the available case law involving the scope of biologic patents."[146]

Panelists and commenters also suggested that the uncertain scope of patent protection was caused by recent rules of the U.S. Patent and Trademark Office (PTO) and Federal Circuit decisions narrowing of the claims to the biologic molecule.[147] A panelist suggested that Federal Circuit decisions and PTO practices have forced patentees to obtain "snapshot" claims that limit the claim scope of the compound patent to the exact amino acid sequence.[148] A panelist suggested that in light of these developments, the PTO is applying the written description requirements in such a way that "it is very difficult to get any kind of scope."[149] Another panelist noted that for biologics, it is much more difficult to establish claims drawn to a broad genus that support current written description and enablement requirements.[150] Additionally, one panelist explained that the PTO recently issued written description guidelines supporting a more narrow interpretation of the written description requirement such that a greater percentage of homology is required in molecules patent claims covering DNA sequences.[151]

Some panelists also discussed the market effects from the Federal Circuit's decisions scaling back the doctrine of equivalents -- a doctrine that allows for a finding of infringement when the infringing product does not fall within the literal scope of the patent claim but is equivalent to the claimed invention.[152] Panelists and commenters suggested that the practical effect of these current trends in patent law portends difficulty for a branded firm to broaden the scope of its patent claims to cover all equivalent products, especially if the biosimilar differs from the branded biologic by a small variation in amino acid sequences.

Another panelist described that when a patent applicant narrows claims during the patent prosecution at the PTO, it has the collateral effect under prosecution history estoppel of surrendering future claims under the doctrine of equivalents that the patentee might try to claim against a biosimilar entrant.[153]

2. Panelists' and Commenters' Arguments that Patent Protection Provides Ample Incentives for Innovation

FOB manufacturers suggested that these arguments were overbroad because of the number and scope of patents that pioneer manufacturers control relating to their biologic products. For example, FOB commenters and panelists explained how the molecule patents claiming branded products would likely be infringed by FOBs.[154] Thus, a "minor and immaterial sequence change is very likely to expose a follow-on biologic to an infringement risk."[155] Another commenter explained that while "smaller biopharma products (such as peptides, fragments and small proteins) may have granted patents covering the full sequence of the product, Amgen's recent success on EPO full sequence claims against Roche and Transkaryotic Therapies (different products and technologies) shows the power in such claims."[156]

Some FOB manufacturers suggested that process patents will likely provide additional protection against infringing products, making it more difficult for FOB manufacturers to design around the patents and obtain FDA approval of an FOB product.[157] For example, one panelist suggested that the pioneer manufacturer may have patented the most commercially viable manufacturing methods and the FOB industry may not be able to devise "another commercially appropriate way to circumvent a process patent."[158] In light of this problem, another commenter suggested that process "patents often provide a level of market protection because the biological origin of their discovery makes them necessary for a production of a product."[159]

Another commenter suggested that additional barriers to FOB entry are created by well-known platform technology patents used in the research, development and manufacture of biopharmaceutical products.[160] This commenter explained that "[t]hese patents are extremely broad and tend to overlap with one another, providing brand biopharmaceuticals with wide-ranging protection over their drug products."[161]

FOB manufacturers suggested that "[b]iologic patents are more likely to obtain patent term extensions under Section 156 [of the Patent Act] due to the long and complex patent prosecutions."[162] These extensions ebb and flow with the PTO's workload.[163] FOB manufacturers also suggested that biologics "are also more likely than chemical drugs to be covered by 'submarine' patents."[164]

D. Patent Protection is Likely to Continue to Provide Strong Incentives for Innovation after Introduction of Follow-On Drug Competition

The patent system has a proven record of protecting and stimulating biotechnology innovation.[165] The introduction of FOB drug products does not alter the relationship of pioneer biologic drug products to the patents protecting them. Pioneer biologic drugs are covered by more and varied patents than small-molecule branded products, including manufacturing and technology platform patents.[166] Patent cases between pioneer manufacturers reveal that patents such as process, manufacturing, and method of use claims can be infringed by a branded competitor.[167] These cases show that the range of patents claiming a biologic product provide a strong assurance that at least one of a biologic drug product's patents will cover an FOB drug product.

There is no evidence that the patents claiming the compound or molecule of pioneer biologic drugs have been designed around more frequently than those claiming small-molecule drug products. There are a variety of ways to draft claims broadly enough to cover the types of drug structure variations expected in follow-on biologics.[168] For example, patent claims reciting the amino acid sequence of a biologic drug compound or molecule can encompass not only the specific sequence, but also a broad genus of structurally and/or functionally related variants through the use of "percent identity claims." An example of a percent identity claim would be 'a protein comprising an amino acid sequence sharing at least 70% identity with the described amino acid sequence.'[169] The PTO's Written Description Guidelines specifically allow the use of percent identity claims.[170] The effect of these claims is that the patent covering the pioneer biologic drug can be broader than the actual product.[171]

Using the example above, an FOB drug product's molecule could differ by up to 30 percent and still infringe the patent protecting the pioneer product.[172]

The scope of drug compound or molecule patents depends on the claim language and patent prosecution. Although it is true that alleged competitors have been found not to infringe drug compound claims because of the way in which the claims were construed,[173] it is equally true that biotechnology drug product claims have been construed so that accused products have been found to infringe even when they have varied from the patentee's corresponding product.[174] For example, a pioneer manufacturer recently obtained a permanent injunction after a finding that its patents were infringed by a competitor that had altered the patented molecule slightly.[175] Other cases are pending as well.[176]

To the extent an FOB manufacturer will attempt to design around a pioneer manufacturer's patent, that effort is to be expected and encouraged. Competing branded manufacturers have been doing just that since the early days of biotechnology patents.[177] The purpose of the required patent disclosures is to assist rival firms to design around patents so that they can develop non-infringing products to compete with the patented discovery and thus spur greater innovation.[178] Of course, FOB manufacturers run the risk that the more their drug molecule differs from the pioneer product's molecule to avoid patent infringement issues, the greater the chance that its product will no longer be "similar" enough to the pioneer product to use the FOB approval process.

Finally, even if the FOB manufacturer were to design around the patents claiming a pioneer biologic drug product and enter prior to patent expiration, the pioneer manufacturer will continue to earn significant revenues after FOB entry. Pioneer manufacturers are likely to retain 70 to 90 percent market share following FOB entry. Moreover, the overall market is likely to expand following FOB entry, thereby diminishing the loss of revenue by the pioneer manufacturer. The effect on the pioneer manufacturer caused by FOB entry is not nearly as great as it is with small-molecule generic drug entry.

In sum, continued reliance on the patent system to stimulate biotechnology innovation is well-justified. This reliance is well-place especially if the FOB abbreviated FDA drug approval process does not contain special regulatory features similar to the ones in Hatch-Waxman that incentivize patent challenges prior to FDA approval of the FOB drug and undermine the ability of the patent to incentivize innovation.

III. Other Considerations do not Support A 12- to 14-Year Branded Exclusivity Period

Commenters and panelists also described the need for, and the likely effects of, a 12- to 14-year branded exclusivity period. The next two sections summarize these views and the need for a 12- to 14-year exclusivity period. The third section provides an analysis of these effects.

A. Panelists' and Commenters' Views on the Likely Effects of a Branded Exclusivity Period

Pioneer manufacturers suggested that a 12- to 14-year branded exclusivity period provides certainty about recoupment when R&D investment decisions are made.[179] Moreover, exclusivity only protects the pioneer manufacturer from the use of its own data by a potential FOB competitor for the length of the exclusivity period.[180]

To calculate the recoupment amount, pioneer manufacturers rely on an economic model (the "Nature model") that calculates the time it takes for a manufacturer to recover fully its investment to develop and commercialize a typical biologic drug.[181] Some commenters have concluded that the Nature model supports a branded exclusivity period between 12.9 and 16.2 years in length.[182]

Pioneer manufacturers suggested that a branded exclusivity period substantially shorter than 14 years would be disastrous for innovation and patients.[183] They suggested that without substantial exclusivity, there will be a decrease in the number of "targets of opportunity" for which FOBs could reference.[184] In addition, R&D would shift away from new treatments for diseases, thus depriving the public of much needed treatments for unmet medical needs, toward 'safer' bets such as new formulations or second generation molecules."[185]

Others suggested that a branded exclusivity period should be similar to the actual amount of time that patented small-molecule products enjoy before generic entry occurs.[186] Under Hatch-Waxman, even though the maximum amount of branded exclusivity is five years, generic entry occurs, on average and depending upon the size of the market, between 11 and 13 years after FDA approval of the branded drug product.[187]

Although one panelist questioned why, if the exclusivity period were short, the pioneer manufacturer could not raise prices to make up for any shortfall in revenue and thus not be any worse off.[188] However, another panelist explained that the "the key driver of prices will be if you're in a market where there's competition or anticipated competition."[189]

By contrast, commenters representing FOB manufacturers suggested that experience under Hatch-Waxman informed their view that a long exclusivity period would lengthen the time between innovations and do little to stimulate innovation. Instead, a 14-year branded exclusivity period may simply reduce the pace of innovation.[190] One commenter predicted that if the branded exclusivity period were that long then branded manufacturers would engage in minor product enhancement strategies which would multiply the costs of expanding monopoly protection.[191] Others noted that long exclusivity periods will eliminate or substantially delay the efforts by FOB manufacturers in making innovations in safety, convenience, cost, access, immunogenicity, interchangeability, or new indications for biologics.[192]

B. Panelists' and Commenters' Views on the Need for Branded Exclusivity to Incentivize Incremental Innovation

The panelists and commenters also examined how the existence and length of an exclusivity period could affect incremental innovation. Pioneer manufacturers explained that there likely would not be incremental innovation without recoupment to recover these investments.[193] In this context, participants used the term "incremental innovation" to refer to actions such as the discovery of a new indication for a previously approved product, or an improved formulation for greater safety or convenience.[194]

A commenter suggested that a 12- to 14- year period of exclusivity is necessary to encourage post-FDA approval research. This commenter explained "that at the time a novel biologic is approved, little may be known of what that drug can do or of what can be achieved in connection with its biological target."[195] Another commenter suggested that without an additional exclusivity period the number of post-approval clinical trials testing new uses of already approved biologic would drastically decrease due to the lack of certainty of an adequate return on investment. For example, instead of anti-cancer biologics being tested in a dozen or more indications in large scale, "Phase IV" clinical trials, no attempt would be made to broaden the use of approved biologic drugs.[196]

Another commenter explained that some extension of exclusivity for the pioneer product is necessary to effectively incentivize the development of new indications for, or other improvements to, existing products. Without such an extension, this commenter predicted that "healthcare practitioners may decide to use the FOB to treat the new indication regardless of whether the FOB was approved for that indication."[197]

Another panelist suggested, however, that there likely would be a trade-off between the length of the initial branded exclusivity period and additional grants of exclusivity for new indications. He suggested that if additional branded exclusivity is granted, that the initial period be kept shorter to encourage the pioneer manufacturer to engage in the post-approval R&D.[198]

C. Likely Competitive Effects of a Branded Exclusivity Period

1. The Innovation Benefits of FOB Competition

As discussed in Chapter 1, FOB manufacturers are likely to seek approval of biosimilar products whose market effects are likely to resemble those of pioneer biologic products rather than small-molecule generic products. Innovation benefits due to branded competition include a race among firms attacking an unmet medical need or investigating a promising therapy that results in increased dissemination of scientific knowledge, and a greater chance of developing a breakthrough product to benefit consumers.[199] The social value of the cumulative effects of incremental innovations can often exceed those of the original breakthrough.[200] These same benefits are likely with entry of FOB products.

Branded competitors also enhance their products to differentiate them from their competitors. This is a common dynamic in competitive markets. Automatic substitution of generic drugs distorts this product enhancement dynamic such that branded manufacturers are incentivized to change their products in minor ways to defeat automatic substitution.[201] These minor changes may not provide clinical or patient benefit. The lack of automatic substitution of FOB products, however, is likely to lessen this distortion in biologic drug markets.

2. Actual Pioneer Drug Manufacturer Exclusivity Can Inform the Length of a Branded Exclusivity Period

The head start that first-in-class branded products already experience against second-in-class products can inform the length of a branded exclusivity period for biologics. A subsequent branded competitor obtains limited benefits from the regulatory approval occasioned by the first-in-class product because its R&D efforts have been proceeding on a parallel path with those of the first-in-class manufacturer.[202] The head start of the first-in-class drug product has decreased over the last three decades as the average lead time of the first-in-class product shrank from 8.2 years during the 1970s to 2.25 years in the 1990s. This limited period of exclusivity for the first-in-class drug product has not dampened R&D incentives and may, in fact, be optimal for rewarding past innovation while allowing competition to incentivize future innovation.[203]

3. FOB Entry is Unlikely to Occur Immediately upon Expiration of a Limited Period of Branded Exclusivity

It is likely that few, if any, biologic products will experience FOB entry immediately upon expiration of a limited period of exclusivity. The generic drug approval process under Hatch-Waxman results in branded manufacturers enjoying approximately 11 to 13 years of *de facto* exclusivity prior to a generic drug entry.[204] This length of market exclusivity occurs despite the incentives within Hatch-Waxman for generic manufacturers to challenge branded patents prior to FDA approval of the generic drug. Indeed, this length of time is attributable mainly to patent protection and patent restoration.[205]

An approach that does not provide incentives to challenge a pioneer product's patents prior to FDA approval is likely to result in a longer period of *de facto* exclusivity than that which occurs under Hatch-Waxman for small-molecule drugs. To the extent patents are at issue, they would be resolved after any branded exclusivity period had expired and FDA approval had been acquired, similar to the way in which branded competitors currently resolve their patent disputes. It is unlikely that FOB manufacturers will expend the substantial resources to develop a biosimilar product and obtain FDA approval if it is likely to run afoul of a pioneer product's patents.

Moreover, expiration of a branded exclusivity period does not mean that FDA approval of an FOB will follow soon thereafter. Pioneer manufacturers are likely to use the citizen petition process to raise safety and efficacy concerns about FOBs, which will delay FOB approvals, as occurred with Omnitrope.[206] Additional delays to FDA approvals of FOB applications would likely occur were FDA required to issue guidance documents, including issuing draft guidance documents, soliciting public comments, and finalizing the guidance documents, before accepting or approving any FOB application for a particular class of branded biologic drugs.[207]

Even if patent litigation were to start following FDA approval, pioneer manufacturers would likely have *de facto* exclusivity for several years after the period ends due to the time it takes to resolve complex patent litigation. The FTC 2002 Generic Drug Study calculated that obtaining a district court resolution of patent issues under the Hatch-Waxman Act took on average 25.5 months and that it took over 12 more months to obtain a court of appeals decision.[208] Given that the patent portfolios for biologic products are likely to include patents owned by third party entities, this time could be substantially extended.[209] Thus, the effect of patent litigation starting after FDA approval of an FOB would delay FOB entry beyond the expiration of a limited branded exclusivity period.

4. Exclusivity Periods Have Been Used When Patent Protection Has Been Insufficient to Incentivize and Reward Innovation

Congress has implemented exclusivity periods to encourage the development of new and innovative drug products when the drug molecule is in the public domain, and therefore not patentable.[210] Similarly, exclusivity periods have been used to incentivize the post-FDA approval clinical trials for new uses of existing drug products. For example, the Hatch-Waxman Act provides a five-year exclusivity period to incentivize the development of new chemical entities. It also provides a three-year exclusivity period for new clinical investigations ("NCI") of small-molecule drugs.[211]

In other instances, Congress has implemented an exclusivity period when market-based pricing has not provided sufficient incentive to develop drug products for target populations.

For example, 6-months of marketing exclusivity periods are awarded upon the showing of safety and effectiveness for children. A seven-year marketing exclusivity period is awarded to manufacturers of drug products that treat diseases affecting less than 200,000 persons in the United States.[212]

Central to each of these exclusivities is a public policy trade-off: a restriction on competition is provided in return for a development of a new drug product or new use of an existing product. A 12- to 14-year exclusivity period, however, departs sharply from this basic trade-off, because it does not spur the creation of a new product or indication. The drug has already been incentivized through patent protection and market-based pricing.

To the extent that there are new biologic molecules that cannot obtain patent protection, an exclusivity period may be warranted. Because there is no evidence about the lack of patentability of new biologic products, nor that market forces have been insufficient to incentivize their development, the Commission has not recommended a length of an exclusivity period.

One benefit of an FOB approval process is that it provides an efficient way to advance scientific progress and commercialization of that scientific innovation. An FOB approval process eliminates unnecessary clinical tests and allows competition to generate better consumer products at lower prices. The potential harm posed by a 12- to 14-year exclusivity period is that firms will direct scarce R&D dollars toward developing low-risk clinical and safety data for drug products with previously proven efficacy rather than toward new inventions to address unmet medical needs. Thus, a new 12- to 14-year branded exclusivity period imperils the efficiency benefits of an FOB approval process in the first place.

In addition, a 12- to 14-year branded exclusivity period could undermine the patent system's disclosure function as pioneer manufacturers rely on trade secrets rather than patents to protect their inventions. Because the patent system requires public disclosure, it promotes the dissemination of scientific and technical information that would not occur but for the grant of a patent. The scientific community can then learn and design around the invention. The ability to design around is prevalent for patent claims covering the formulation or dosage of drug products, product-by-process claims, and process claims – all of which currently protect pioneer biologic products. To the extent that the branded exclusivity period replaces the need for the patent, the scientific community loses the disclosure of inventions that occurs when patents are granted and published, and innovation could be harmed.

D. The Nature Model Fails to Inform Reliably the Length of a Branded Exclusivity Period

Pioneer manufacturers have developed the Nature model to show that the optimal length of branded exclusivity should be approximately 14 years. The Nature model, as currently presented, contains numerous methodological and conceptual weaknesses that render its results too imprecise and non-robust to inform discussions about the length of an exclusivity period. A model that balances the benefits of FOB competition with the costs of potentially forsaking marginal branded drug development projects would be more informative than the Nature model's approach.[213]

Appendix A further explains and evaluates the assumptions underlying the Nature model. A brief summary of the problems includes:

- **Imprecision:** The estimates of costs and revenues used in the model are based on extremely small samples of drug products and are likely imprecise.
- **Inelastic Demand:** Most versions of the model currentlyassume that the overall quantity of the drug produced and sold will not expand with FOB entry although they assume that FOB entry will lead to lower prices.
- **Internal Inconsistency:** The ad hoc assumptions about the branded manufacturers' price decrease and market share decline following FOB entry are not necessarily consistent with the likely market dynamics of FOB competition.
- **Excessive Aggregation:**
  - The revenue estimates do not distinguish between the original and subsequent indications and formulations, so an independent analyst cannot modify the framework to calculate the break-even point for just the original indication and formulation.
  - The model is based on a portfolio of biologic drugs that includes blockbuster drugs as well as drugs with relatively less in sales and profit. The use of an average revenue stream likely produces an exclusivity period that overprotects the top-selling drugs which are the only drugs likely to face FOB entry when the branded exclusivity period ends.
- **Non-Robustness:** The model's results are extremely sensitive to small changes in the cost of capital[214] and other assumptions.

3. Competitive Effects of a Pre-Approval Patent Resolution Process

I. Introduction

Patent protection fuels the biotechnology industry's R&D engine.[215] As discussed in Chapter 2, patent protection and market-based pricing enable biotechnology firms to increase their expected profits from investments in R&D, thus fostering innovation that would not occur without patents' exclusionary rights.[216]

Special procedures to provide an early start to resolving patent disputes between pioneer and FOB manufacturers prior to FDA approval of the FOB product are unlikely to be successful to facilitate FOB entry. Although special procedures govern patent litigation between branded and generic competitors over small-molecule drug products, these procedures are the exception, not the norm. In every industry, including the biotechnology industry, competing firms have engaged in patent litigation in which the patent holder initiates infringement litigation or the alleged infringer seeks a declaratory judgment of non-infringement or invalidity. In the biotechnology industry, this process usually begins following FDA approval of the competing drug product.[217]

The special procedures for small-molecule drugs were designed in 1984 to address the issue of "judgment proof" generic defendants. In this context, the profits of the alleged infringer (the generic entrant) are substantially less than the loss of profits by the branded product manufacturer, because of the substantial price differences between branded and generic products. Consequently, generic entrants in small-molecule drug markets are unlikely to be able to satisfy a potential treble damage award for infringing the branded manufacturer's patents.

This chapter explains that FOB entrants will not be similarly judgment proof. FOB entrants are not expected to offer the deep discounts seen in small-molecule drug competition. Rather, FOB entry is likely to resemble the market impact of entry by subsequent branded entrants. An FOB manufacturer is likely to introduce its FOB product at prices 10 to 30 percent lower than the pioneer manufacturer's price. Because FOB entrants will earn greater profits, and will be able to satisfy potential damage awards, the market dynamics of FOB competition do not justify creation of a special regulatory system to protect pioneer manufacturers from judgment-proof defendants.

Although FOB market entry would be eased if an FOB manufacturer had complete certainty as to whether its product infringed the pioneer product's patents, a pre-FDA approval patent resolution process is unlikely to provide greater certainty than use of existing statutory patent resolution mechanisms.[218] A special pre-approval patent resolution process is not likely to succeed in raising and resolving all pertinent patent issues prior to FDA approval. Patents claiming the pioneer product may issue after a pre-approval process has begun and/or after FDA approval. In either situation, the FOB manufacturer will need to resolve these later-issued patents before commercial marketing. The FOB manufacturer's application and product also may change during the approval process such that by starting patent litigation prior to FDA approval would not ensure earlier resolution. Moreover, without a mechanism to enforce the rules of a pre-approval resolution process, there is no guarantee that litigation that is started prior to FDA approval will end earlier. Incorporating a pre-approval patent resolution process into a 12- to 14-year branded exclusivity period is unlikely to mitigate these problems.

Based on the experience under Hatch-Waxman, a pre-approval patent resolution process also is likely to lead to consumer harm, including the facilitation of anticompetitive conduct that defeats the purpose of starting the patent litigation early. In the Hatch-Waxman context, branded manufacturers have used the pre-approval patent regulations to delay generic entry. In addition, generic and branded competitors have entered into "pay-for-delay" patent settlements that delay entry, not encourage it. It is likely that a pre-approval patent resolution process in the FOB context could facilitate collusive agreements and/or provide the pioneer biologic drug manufacturer with competitively sensitive information about a significant potential competitor to which it otherwise would not have access.

II. The Likely Effects of an FOB Pre-Approval Patent Resolution Process

A. Background on the Hatch-Waxman Pre-Approval Patent Resolution Process

Current law exempts the FOB manufacturer from patent infringement liability for work directed towards petitioning the FDA for product approval.[219] To be liable for infringement, an FOB manufacturer must take steps separate and apart from seeking FDA approval, in essence, it must import, make, use, sell, or offer to sell its product.[220] The FOB manufacturer, however, is unlikely to take these steps until it receives FDA approval. Consequently, to have patent litigation begin *before* FDA approval of the FOB, Congress must create an "artificial act of patent infringement" and a mechanism to resolve subsequent patent litigation (a "pre-approval patent resolution process").

Hatch-Waxman established special procedures to incentivize generic small-molecule drug manufacturers to challenge invalid or narrow patents on branded products. These procedures allowed the patent resolution process to run concurrently with the FDA regulatory approval process. Of course, to the extent a generic applicant seeks entry on the day the last patent claiming the branded drug product expires, these procedures are not utilized.[221]

To effectuate the pre-approval patent resolution process, Hatch-Waxman requires branded manufacturers to list certain patents claiming the branded drug product in the FDA's Orange Book. A generic applicant is then required to certify whether it seeks FDA approval prior to the expiration of any of the patents listed in the Orange Book that covers the referenced branded product. If it does, the generic company must provide notice to patent holders and the branded product manufacturer. The notice must include a detailed statement of the factual and legal basis supporting the applicant's assertion that the listed patents are invalid or not infringed.

To incentivize early pre-approval litigation and resolution, if the branded manufacturer brings infringement litigation within 45 days from notice, the FDA cannot approve the generic application for 30 months.[222] If the branded manufacturer does not initiate litigation within 45 days, the FDA is free to approve the generic application in its normal course.

B. Commenters' and Panelists' views on the Likely Effects of an FOB Pre-Approval Patent Resolution Process

Panelists and commenters described the likely benefits of developing a pre-approval patent resolution process derived from the Hatch-Waxman Act. Panelists suggested patent infringement certainty would enhance their drug development activities.[223] They also suggested that a pre-approval patent resolution process is likely to preclude FOB at-risk launches, which occur when a company launches its product without knowing if all product-related patent issues are resolved.[224] Other panelists predicted that certainty is likely to attract venture capital resources,[225] and suggested that smaller companies may not be in a position to launch-at-risk because they are unlikely to attract investment funds without certainty.[226] Another panelist predicted that without a pre-approval resolution process, pioneer manufacturers would not be able to enforce injunctive relief against an FOB entrant and that this would lead to compulsory licensing of patents rather than removal of the product from market.[227]

Panelists and commenters described the likely effects of linking FDA approval to the outcome of patent litigation, as it is done under the Hatch-Waxman Act. Panelists representing pioneer manufacturers explained that if a court finds a pioneer product's patent to be valid and infringed, the FDA should not approve the infringing FOB product until the patent expires.[228] Another suggested that tying FDA approval to patent resolution would "keep follow-on biologics that infringe a patent off the market by preventing final FDA approval until patent expiry."[229]

One panelist and commenter representing a potential FOB entrant suggested that linkage is unnecessary because existing patent law provides for robust protection of patent rights against infringement and is sufficient to deter inappropriate entry.[230] This panelist explained that launching-at-risk currently is the norm in the biotech industry.[231] He also explained that the generic small-molecule industry is the only industry that has an artificial act of infringement, and "that was a result of the state of the industry in 1984, and we don't believe [it] is required with the state of the industry in 2008."[232]

Another panelist noted that linkage may be unworkable because biologic patent portfolios often include patents that have been licensed to third parties. Infringing one of these patents may not lead to a permanent injunction, and thus should not preclude FDA approval.[233] Another panelist added that the patent holder may obtain an injunction notwithstanding the fact that they have licensed the patent to other parties.[234]

Participants also described the likely unintended consequences of a pre-approval patent resolution process, which include: delay of FOB entry, distortions to the parties' incentives during the process, and increased costs.[235] For example, one panelist suggested that the process can cause unintended delay, noting that for small-molecule drugs, Hatch- Waxman has led to "serial litigation."[236] Another panelist focused on the likelihood of wasteful litigation.[237]

Panelists also described how a pre-approval resolution process will distort parties' incentives. One panelist explained that the process will "create bounties on valid patents" by producing an incentive for an FOB to challenge patents before it has shown it can develop an approvable drug.[238] By starting the process early, that is prior to FDA approval, parties are encouraged to bring multiple litigations, where, if they brought litigation at the end of the process, there may be greater incentive to raise only the strongest patents.[239] Commenters also explained how if litigation begins too early, the FOB application and product also may change during the approval process such that an early start to litigation prior to FDA approval would not ensure an earlier resolution. Rather, an early start to litigation would lead to additional litigation upon finalization of the application and the FOB product.[240] Panelists also described how unnecessarily-early litigation processes will increase pioneer and FOB costs, explaining that a pre-approval patent resolution process likely will "bring[] on expensive litigation costs earlier when you might not want to do that."[241]

C. Analysis of the Likely Effects of a Pre-Approval Patent Resolution Process

1. The Likely Market Impact of FOB Drug Entry Does Not Warrant a Special Pre-Approval Patent Resolution Process

The justification for special procedures akin to those in Hatch-Waxman for small-molecule drugs depends upon the context in which FOB competition is likely to proceed. As discussed in Chapter 1, FOB competitors are likely to seek approval of biosimilar products. The competitive dynamics of biosimilar entry are likely to resemble entry by a branded drug product, in which FOB competitors introduce their products at discounts between 10 and 30 percent of the pioneer products' price. This effect contrasts with the 80 percent discounts that occur with entry of multiple small-molecule generic products.[242] The competition prompted by biosimilar entry is unlikely to move more than 10 to 30 percent market share away from the pioneer manufacturer. This market share movement is substantially less than the market share gain that small-molecule generic drugs obtain due to state substitution of generic drugs. Because of smaller discounts and smaller market share, the FOB entrant is unlikely to be judgment-proof and thus able to pay any possible damages resulting from infringing a pioneer product's patents.

Because FOB entrants are likely to mimic the market effects of another branded product, the FOB and pioneer manufacturer can avail themselves of the existing patent litigation procedures that apply to every industry, except generic small-molecule drugs. Biologic drug manufacturers have successfully used this process to resolve patent litigation for decades.[243] It is the same process all patent holders use to resolve claims of infringement or validity – the patent holder initiates infringement litigation after the FDA has approved the potentially infringing drug product, or the alleged infringer seeks a declaratory judgment of non-infringement or invalidity. For example, the Supreme Court has addressed biotechnology patent disputes in *Merck KGaA v. Integra Lifesciences I, Ltd.* and *MedImmune Inc. v. Genentech, Inc.*[244] Lower courts also have addressed patent infringement litigation against competing products that had obtained FDA approval but had not yet been marketed.[245]

In other words, the patent holder can use existing court remedies to enforce its patent rights against an FOB, without developing special procedures that condition FDA approval on the outcome of patent litigation. Although the lack of a pre-approval patent resolution process increases the potential for at-risk launches by an FOB, a profit-maximizing FOB manufacturer is unlikely to enter the market "at-risk" if it believes it will be liable for substantial infringement damages, cause physician and patient confusion, or harm its reputation as a reliable FOB drug manufacturer.

2. A Pre-Approval Patent Resolution Process is Unlikely to Provide Certainty and is Likely to Disrupt Innovation Incentives

A pre-approval patent resolution process is unlikely to achieve the certainty goals desired by pioneer and FOB manufacturers for three reasons. First, pioneer manufacturers with vulnerable patents have no incentive to have their patents invalidated or held not infringed by the FOB drug, especially if such a determination were to come several years before patent expiration or before FOB entry was imminent. In other situations, a pioneer manufacturer or third party may license its patents to other market participants where the license creates a revenue stream for the patent holder. If these patents are deemed invalid, the patent holder loses this revenue stream.

Experience under Hatch-Waxman shows that profit-maximizing manufacturers are likely to use a pre-approval regulatory process to delay a final court decision.[246] In other instances, pioneer manufacturers may seek to bring suit in a judicial district with a history or reputation of slow-moving proceedings or they may fail to participate in the process, thus requiring the generic firm to bring a declaratory judgment action.[247] Pioneer manufacturers also may attempt to hold back relevant patents during this pre-approval process if the regulations are subject to interpretation or the penalty for violating the rules provides an insufficient deterrent. These tactics can succeed because the FOB product has not been approved and the FOB manufacturer is unable to threaten market entry to further the process along.

By contrast, if litigation were to begin post-approval, the way in which branded biologic competitors resolve patent issues currently, a patent holder is likely to assert its strongest patents to keep the FOB product off the market. This process naturally focuses litigation on the strongest patents, and reduces unnecessary pre-approval litigation regarding patents that may not be asserted after FDA approval.[248]

Moreover, without special pre-approval processes, there would be no need to change the declaratory judgment standards or rules. Even if the pioneer manufacturer did not initiate patent infringement litigation, the newly-approved FOB entrant would have standing to seek a declaration that the pioneer manufacturer's patents are invalid or not infringed.

Second, a start to litigation prior to FDA approval does not guarantee that patent issues will be resolved earlier than if litigation begins after FDA approval. Patents claiming the pioneer product may issue after a pre-approval process has begun, but before FOB approval. Patents also may issue after FOB approval. In either situation, the pioneer manufacturer, or third party, will need to bring additional litigation to enforce these later-issued patents, removing the certainty sought by the parties. The FOB's application and product also may change during the approval process, such that early patent litigation would no longer apply to the approved product. The litigation would be about a "moving target." Moreover, without an enforcement provision, even with a pre-approval process there is no guarantee that litigation will begin pre-approval. Until the FOB product is approved, patent infringement litigation may be premature.

Third, patent litigation under Hatch-Waxman shows that a pre-approval process is likely to invite numerous patent challenges. In the Hatch-Waxman context, nearly every branded drug faces a pre-approval patent challenge.[249] Similarly, in the FOB context, a pre-approval patent resolution process may incentivize FOB manufacturers to challenge all of a pioneer product's patents in hope of exposing and exploiting weaknesses in the patent portfolio. In contrast, the absence of a pre-approval patent resolution process is likely to incentivize FOB manufacturers to direct their product development resources to those areas in which the pioneer product's patents are likely to be invalid or not infringed.

3. A Pre-Approval Process is Unlikely to be Workable and is Likely to Cause Harm

At a minimum, a pre-approval process must include two components: (1) notification requirements, including when notification begins; and (2) identification of patents to be litigated in the pre-approval period, which could include only "necessary" patents. The following sections describe how these procedures are unnecessary, could lead to anticompetitive outcomes, and defeat the purpose of a pre-approval process to obtain early resolution of potential patent infringement issues.

In addition, strong enforcement of the governing regulations for the pre-approval process will be necessary to deter abuse by the participants that seek to use the process to obtain competitive advantages. It is likely that a self-policing process will not work and that FDA will be asked to referee the process, much like it has been forced to do with the Hatch-Waxman process.

a. Notice Provisions are Unnecessary and Could Raise Anticompetitive Concerns

To be effective, a pre-approval patent resolution process will need to incorporate two major types of notice: (1) the pioneer manufacturer will need to provide notice to potential FOB manufacturers of patent claims covering its pioneer products; and (2) the FOB manufacturer will need to provide notice to the pioneer manufacturer of its FDA application.[250]

(1) Patents Claiming the Pioneer Drug Product are Publicly Available

Although the first type of notice is likely to help the FOB identify which patent claims its product may infringe, it is unnecessary given that granted patents and post-2000 patent applications are published by the PTO.[251] FOB manufacturers can use existing databases to perform a patent search, as companies in many industries do, to determine patent claims that its product may infringe.[252] This search would apply to patents owned by the pioneer manufacturer and any applicable third parties.

In addition, the Patent Act currently requires patent holders to provide notice of potentially infringed patents. A patentee cannot recover damages for infringement until it (1) marks the product; or (2) provides the alleged infringer with actual notice of the infringement.[253] If the product, or its packaging, is not physically marked with applicable patent numbers, then the patentee can give notice either by sending a warning letter to the alleged infringer, or by bringing a suit for infringement.[254] The notice must identify the patent(s) and specifically allege infringement.[255] For process patents only, an additional "request for disclosure" applies. Before it sells its product, a competitor or potential competitor may request a patent holder to produce all process patents that the patent holder believes could be infringed by the competitor's product.[256] These notice procedures would apply to follow-on biologic drugs even if there were no pre-approval patent resolution process.

A patent listing system also is likely to lead to anticompetitive unintended consequences. For example, Hatch-Waxman's notice provision led to the delay of generic entry until the notice provisions were amended by the Act's 2003 Amendments. As discussed above, a branded manufacturer must list certain patents in the FDA's Orange Book.[257] The generic then files a certification regarding each patent. If the branded manufacturer then brings an infringement action within 45 days, FDA approval of the ANDA automatically is stayed for 30 months.[258]

Over time, branded manufacturers began successively to list later-issued patents in the Orange Book. A number of these later-listed patents did not meet the FDA's requirements for listing patents in the Orange Book and were subsequently found to be invalid or not infringed.[259] This strategy allowed the branded manufacturer to obtain additional 30-month stays delaying FDA approval of generic drugs. Congress remedied this problem in the Medicare Modernization Act by limiting branded drug companies to a single 30-month stay, but only after consumers lost substantial competition from generic drugs during the periods of these "stacked" 30-month stays.

(2) Notice of the FOB's Application Raises Competitive Concerns

The FOB manufacturer's notice to the pioneer manufacturer of its FDA application and additional manufacturing information raises two concerns – one administrative and one anticompetitive. First, there is a difficulty in determining to whom the notice should be provided. Biologic drug patents implicate more than the pioneer manufacturer; they also implicate universities and third parties. If notice is provided only to the pioneer manufacturer, an early start to patent resolution may not involve all of the relevant parties and patents, thus defeating the purpose of the pre-approval process. Conversely, if notice is provided to all parties, it may overly complicate the process and deter early resolution. Indeed, one panelist suggested "[t]he whole issue of notice should be as simple as possible, but some of the issues are more complex than we see even in the more complex drug situations."[260] This complexity is likely to reduce the effectiveness of a notice requirement and raise questions over whether sufficient notice had been provided in a timely manner.

In addition, requiring the FOB manufacturer to provide a detailed description of the product and manufacturing processes to the pioneer manufacturer and other third parties could facilitate anticompetitive conduct.[261] As discussed in Chapter 1, the FOB manufacturer is likely to compete against the pioneer product with a similar, but not identical, product. The FOB product could be an improvement over the pioneer product in terms of reduced dosing, increased effectiveness, or fewer side effects. In other cases, the FOB product manufacturer may have discovered a way to manufacture an FOB product more efficiently than the pioneer manufacturer. In either scenario, the firms are likely to be significant rivals and engage in head-to-head competition.

Forced sharing of information between rivals about the timing and content of the FOB's application and manufacturing processes (and other related matters) could facilitate collusion.[262] For example, this information could facilitate agreements to delay entry, allocate markets, or fix prices. Experience under Hatch-Waxman has shown that generic and branded competitors have entered into "pay-for-delay" patent settlements that delay entry. In other situations, the anticompetitive harm could stem from providing the pioneer manufacturer with competitively sensitive information that it otherwise would not be able to obtain. The pioneer manufacturer may then have an opportunity to act on this information prior to the approval of the FOB and thus, can pre-empt the innovation and price competition that is likely to occur with FOB entry. This harm is lessened, although not eliminated, with patent litigation after FDA approval because the FOB can enter quickly and blunt any harm that could be caused by a sharing of competitively significant information.

Moreover, sharing this type of information may be unnecessary to the extent that the FOB manufacturer claims that the pioneer manufacturer's patents are invalid. In these situations, the pioneer manufacturer has no need for detailed information relating to the FOB's method of manufacturing.[263]

Confidentiality provisions that limit access to information about the FOB product to persons involved in the pre-approval patent litigation process are likely to be ineffective to safeguard against these potential anticompetitive harms. Biotechnology patent litigation is a complex endeavor that requires not only patent attorneys, but medical, scientific, manufacturing, and business personnel with intimate knowledge about the pioneer product's patent claims. There is no way to cordon off a patent infringement analysis regarding the FOB drug product from the personnel that know the most about the pioneer product.[264]

b. Identification of a Subset of Patents to Resolve During the Pre-Approval Patent Resolution Process Defeats the Purpose of a Pre-Approval Resolution Process

A pre-approval patent resolution process will need to account for the broad range of patents claiming a pioneer product and their multiple owners. As described in Chapter 2, the types of patents that are likely to claim the biologic product include compound or molecule patents, method of treatment, formulation and dosage form patents, and manufacturing process and technology platform patents. Due to the nature of biologic drugs, these portfolios may include patents owned by the pioneer manufacturer,[265] as well as third-party owned patents that are licensed either exclusively or non-exclusively to the pioneer manufacturer.

Panelists representing pioneer manufacturers proposed that a pre-approval patent resolution process resolve all of the patents claiming a pioneer product.[266] Another panelist noted that the process should include third-party patents, reasoning that if the patent resolution process does not cover third-party patents, then the generic will be susceptible to launch-at-risk on those patents.[267]

As a counterpoint, panelists said that litigating all patents potentially infringed by the FOB product could lead to delay. If the process is not tailored, litigation costs could outweigh benefits for some FOB companies.[268] Commenters representing FOB manufacturers explained that the process should be limited to "necessary" patents (*i.e.*, patents that would most likely prevent the FOB from entering the market). One commenter explained that "it is in the generic's interest to immediately litigate only those patents that would prevent the generic company from launching until questions of validity, enforceability, or infringement are resolved. Litigation on all remaining patents would take place after the generic product actually enters the market."[269]

It is unclear how a regulatory process could determine which patents are "necessary." Resolving infringement issues for a subset of "necessary" patents may streamline that particular litigation, but it is likely to lead to future uncertainty. A patent not deemed "necessary" does not mean that the patent is invalid and/or not infringed by the FOB manufacturer. The patent holder can still assert these "unnecessary" patents following FDA approval of the FOB. Retaining these later "unnecessary" patents is unlikely to create the certainty that a pre-approval patent resolution process is intended to create. As noted above, additional patents that block FOB entry may issue after the pre-approval process has started, thus complicating identification of "necessary" patents and frustrating the overall objective of the pre-approval process to obtain certainty regarding patent infringement issues.

Furthermore, a two-tier resolution system whereby some patents are "necessary" and litigated pre-approval, while others must wait until after FDA approval, will likely create additional litigation regarding the determination of "necessary" versus "unnecessary" patents. Such litigation will create additional costs detrimental to consumer welfare. Moreover, as discussed above, Hatch-Waxman created the incentive for the branded manufacturer to "stack" patent notification to obtain multiple 30-month stays. Here, too, limiting the process to a subset of patents may create incentives for the brand to withhold certain "necessary" patents to retain their rights after FDA approval of the FOB product.

c. Enforcement Provisions May Harm Innovation and Competition

An enforcement provision is likely to be necessary to ensure that the notice and patent identification requirements are adhered to during the pre-approval resolution process. Without such an enforcement mechanism, the pre-approval process is unlikely to be adhered to, and likely to cause unintended consequences that delay FOB entry.

Several panelists and commenters suggested that a "sue-or-lose" provision in which a pioneer manufacturer, or third party, could lose its patent enforcement rights if it did not participate in the patent resolution process, is necessary to ensure the integrity of the pre-approval patent resolution process.[270] One panelist said that without a penalty provision, the rules likely will not be enforced, noting that if the pioneer manufacturer holds back patents until the end of the exclusivity period, or launch, then any likely effect of early resolution will not be achieved.[271]

Panelists representing pioneer manufacturers opposed a sue-or-lose provision or a provision in which damages are limited for lack of participation in the process.[272] One commenter said that it likely would lead to gaming, "the patent owner would be forced to decide whether to sue based on the information it obtained from the FOB applicant. That applicant, in turn, would have an incentive to convince the patent owner not to bring a suit."[273] Another panelist explained that a sue-or-lose provision would take away a valuable property right from the patent holder for failure to comply with a regulatory obligation.[274]

Based on experience with the patent resolution process in Hatch-Waxman, examples of the need for enforcement could include:

- If the patent holder's notice fails to include all of the patents or all of the "necessary" patents claiming the pioneer product;
- If the patent holder fails to update the notice to include patents issued after the patent litigation has begun;
- If the FOB applicant fails to provide a sufficiently detailed description of the FOB product, its method of manufacture, or the materials included in the manufacturing; or
- If either the pioneer manufacturer or FOB applicant fails to provide information in a timely manner.

Experience under Hatch-Waxman also demonstrates that the FDA will be pulled into these disputes and asked to resolve substantive patent issues to enforce the rules in any patent resolution process.[275] Under Hatch-Waxman, the FDA has been asked to determine whether patents are correctly listed in the Orange Book. The FDA has consistently maintained that it does not have the patent expertise to do so.[276] Its resources are likely to be best deployed in examining the safety and effectiveness of FOBs, not in policing a patent resolution process for which it has little experience and expertise.

If the FDA were not involved, an enforcement provision could be designed so that if a party did not sue under the patent resolution process in a timely manner, it would lose its rights to later enforce the patent under provision of the Patent Act. This provision is beneficial because in order for the process to have integrity, there must be a mechanism to compel parties to participate. On the other hand, it is likely to be an unduly harsh remedy in the face of uncertainty as to a determination of "necessary" patents to include in a notice or the extent of detail in the FOB applicant's notice describing its product and its method of manufacturing. Such a remedy may also unnecessarily affect the patent holder's right to assert the patent in unrelated contexts. Resolving these uncertainties (and others) is likely to delay resolution of patent issues and, thus, defeat the purpose of a pre-approval patent resolution process.[277]

In sum, although there may be legitimate issues about invalid or not infringed patents blocking FOB drug entry, these issues are best handled post-FDA approval when the parties' incentives are not distorted by a pre-approval process. In other words, post-FDA approval, the FOB manufacturer will seek to begin commercial marketing and the pioneer manufacturer will seek to obtain a preliminary injunction to block FOB drug entry.

4. Likely Competitive Effects of a Market Exclusivity Period for Follow-On Biologics

The Hatch-Waxman Act provides a 180-day marketing exclusivity period to the first generic drug applicant that seeks FDA approval prior to the expiration of patents relating to the branded drug product.[278] No other generic manufacturer may obtain FDA approval to market its product until the first generic has sold its product for 180 days or has forfeited the exclusivity period.

The 180-day exclusivity period incentivizes generic manufacturers to challenge the patents claiming a branded drug product. One court has explained that the 180-day exclusivity rewards the first generic applicant for the expense and effort involved with patent challenges.[279] A court finding of patent invalidity benefits not only the challenger, but also subsequent generic applicants whose entry is no longer blocked by the patent. Thus, the 180-day marketing exclusivity period prevents immediate free-riding by subsequent generic applicants on a favorable outcome that results from the first applicant's patent challenge. As subsequent generic firms enter, generic prices can drop to 80 percent off the branded price, depending upon the number of entrants.[280] The exclusivity period permits the first generic entrant to recoup its patent litigation costs before the substantial price drop caused by multiple generic entrants.

This chapter summarizes the commenters and panelists views on the need for, and the likely effects of, providing FOB manufacturers with incentives to develop their products by restricting entry of competing products during an FOB exclusivity period. It then explains that an exclusivity period is unnecessary to encourage the development and marketing of biosimilar products. Biosimilar products are likely to earn substantial profits without regulatory exclusivity periods. Moreover, European and U.S. experience with biosimilars shows that sufficient profit incentives already exist to encourage biosimilar entry.

An exclusivity period is likely to be unnecessary to encourage the development of interchangeable biosimilar drug products because potential market opportunities appear robust. The competitive dynamics that justified the 180-day exclusivity period for small-molecule generic drugs are unlikely to be present with the entry of interchangeable biosimilar drugs.

It also is unclear that an exclusivity period will successfully incentivize a manufacturer of a biosimilar product to develop an interchangeable FOB product. Biosimilar manufacturers are likely to make this additional investment based on a consideration of, among other things: the cost, expected prices, capacity constraints, and the extent and effect of state substitution laws. This situation contrasts significantly with small-molecule generic competition seen under Hatch-Waxman in which generic manufacturers enter initially with an interchangeable product. Unlike FOB manufacturers, generic manufacturers do not market a "similar" product first and replace it with an "interchangeable" product later.

Not only do market dynamics counsel against an FOB exclusivity period, but the anticompetitive delay in entry evidenced in small-molecule generic drug markets is likely to be repeated if an exclusivity provision for interchangeable FOBs is implemented.[281] The current 180-day exclusivity period exacerbates the problem of "pay-for-delay" settlements that prevent generic entry.[282]

Awarding an FOB exclusivity period on a "first-to-approve" rather than a "first-to-file" basis does not lessen the potential harm. These anticompetitive consequences are likely to result if the period can be extended, the period does not run immediately upon its award, or if a firm has the ability to delay triggering the running of the period through, for example, a patent settlement, acquisition, merger, or agreement.[283]

I. Necessity of an Exclusivity Period to Encourage Development of Follow-on Biologic Drugs

The question arises whether an FOB manufacturer needs an incentive beyond market-based pricing to develop an interchangeable FOB drug, such as a limit on when subsequent interchangeable FOB drug entry can occur (an "FOB exclusivity period"). This limitation would allow the first interchangeable FOB manufacturer to recoup its development expenses.[284] One commenter indicated that "most companies contemplating biogenerics will be reluctant to invest the significant resources required to determine interchangeability if there is no possibility for recouping the costs that come with patent challenges."[285]

By contrast, other commenters and panelists suggested that there was no need for an FOB exclusivity period because potential market profits would provide sufficient incentives to enter with a follow-on product.[286] Another commenter explained that if the market did not provide sufficient incentives on its own, an FOB exclusivity period would not do so either. An applicant would already have the assurance of *de facto* exclusivity, because there would not likely be a second or subsequent entrant, and *de jure* exclusivity would add nothing to the economic calculus.[287]

One potential FOB entrant explained that if a company invests a huge amount of money developing FOB products, it is unwise to put up further barriers in the form of exclusivity granted to other FOBs against its ability to get a return on investment.[288] Others panelists stressed that an additional incentive to foster FOB entry is no longer needed because the environment in 2008 is much different that it was in 1984, when Hatch-Waxman was enacted and there were no established generic competitors, thus an incentive was necessary to jump start the industry.[289]

Other panelists questioned why an FOB exclusivity period was needed in the United States, the largest drug market in the world, when follow-on drug manufacturers in Europe have not needed exclusivity to incentivize biosimilar entry. One panelist noted that European regulatory structure does not provide market exclusivity for biosimilars or "for any generic of any kind, including small molecule."[290] This panelist also concluded that because many generic companies do not make the 180-day period the cornerstone of their business model, the 180-day marketing exclusivity period is not necessary to encourage entry.[291] A commenter also noted that the large number of biosimilar products under development in Europe, where no market exclusivity is provided for biosimilar products, indicates that market exclusivity in the United States may be unnecessary.[292]

II. Marketing Exclusivity Limited to Interchangeable Follow-On Biologics

Panelists suggested that any regulatory exclusivity period be limited to interchangeable FOBs (*e.g.*, biogenerics), which may cost more to develop than biosimilar drug products.[293] One panelist indicated that demonstrating interchangeability may require clinical trials. The trials will be complicated and expensive if there are multiple interchangeable products.[294] A commenter suggested that a "short period of exclusivity for the first to market could provide an incentive to companies entering the biogeneric market; however, companies will not likely rely on winning exclusivity to invest in the products because the development time and investment for biogenerics is so great."[295]

Some panelists suggested that there are likely to be few, if any, interchangeable FOB entrants because of the additional expenses to develop and obtain approval of interchangeable products.[296] In contrast, another panelist predicted that the *availability* of an FDA approval process for interchangeability would prompt the development of the necessary analytics needed to prove interchangeability.[297]

If there were an FOB exclusivity period, panelists described how experience under Hatch-Waxman provided insights into how best to structure the exclusivity to avoid unintended anticompetitive effects that delayed entry. One panelist suggested that an FOB exclusivity period should restrict only other interchangeable FOBs, not other biosimilars, from coming to market during that period of time.[298]

One commenter posited that the Hatch-Waxman 180-day marketing period was designed to incentivize generic drug applicants to engage in patent litigation because of the concern that other generic drug applicants would free-ride on this litigation investment.[299] One panelist suggested that placing a bounty system on intellectual property rights through the awarding of marketing exclusivity for patent challenge is not in the public interest.[300] Other commenters suggested that FOB exclusivity "be based on product approval rather than patent challenge" such that it does not create a "perverse incentive to challenge the innovator's patent early and often, regardless of the merit of the challenge."[301]

III. Analysis of Likely Competitive Effects of an FOB Exclusivity Period

An FOB exclusivity period is unnecessary to encourage the development and marketing of biosimilar products. Market forces to incentivize the development of these products appear robust. Indeed, several panelists and commenters noted that the multi-billion dollar size of the market opportunities, the European experience of HGH and EPO biosimilar entrants, and the U.S. EPO biosimilars provide strong evidence to predict that regulatory incentives are unnecessary to encourage biosimilar products in the United States.[302] Moreover, they are likely to face less competition than small-molecule generic drug manufacturers because of the high entry costs.

An FOB exclusivity period also is unlikely to be necessary to encourage the development of interchangeable biosimilar drug products for several reasons. First, the conditions that justified the 180-day exclusivity period for small-molecule generic drugs under Hatch-Waxman are unlikely to be present. Interchangeable FOB drug prices are unlikely to fall as much (either in real terms or as a percentage of the pioneer product's price) as they do when multiple small-molecule generic drugs enter the market. In the small-molecule generic drug context, the first generic entrant is able to recoup its patent litigation costs before entry of additional generic drugs. Additional generic entry substantially decreases the generic price, in some cases, up to 80 percent off the referenced product's price. It is likely, however, that few interchangeable FOB entrants will enter the market, and prices will not fall as much as they do following small-molecule generic drug entry.

Second, it is unclear whether subsequent interchangeable entrants would be able to "free-ride" on the first interchangeable's FDA approval or patent litigation expense and thus enter the market once the first interchangeable product is approved. It is expected that FDA approval of interchangeable products (and accompanying patent litigation) is likely to be more complicated than generic drug approval. Unlike generic small-molecule drugs where several generic drug products often await FDA approval once a patent expires or is found invalid or not infringed, this complexity is likely to diminish the prospect that a "queue" of interchangeables will be ready for approval once the first interchangeable product is approved. Thus, the circumstances that justified a 180-day marketing exclusivity period for generic drugs are unlikely to be present for interchangeable FOB drug products.

Third, it is uncertain that cost will justify an FOB exclusivity period. It may not cost substantially more to show that a biosimilar product is interchangeable with the referenced branded product than an initial finding of biosimilarity. If technology advances such that it is relatively inexpensive to determine interchangeability, an exclusivity period is unnecessary.

Fourth, it is unclear that an FOB exclusivity period will successfully incentivize a manufacturer of a biosimilar product to develop an interchangeable FOB product. Biosimilar manufacturers are likely to make this additional investment based on a consideration of, among other things, the cost, expected prices, capacity constraints, and the extent and effect of state substitution laws. This situation contrasts significantly with small-molecule generic competition seen under Hatch-Waxman in which generic manufacturers enter initially with an interchangeable product. Unlike FOB manufacturers, generic manufacturers do not market a "similar" product first and replace it with an "interchangeable" product later.

Not only is an FOB exclusivity period not justified by market conditions but the delay in generic entry evidenced in small-molecule generic drug markets is likely to be repeated in biologics markets if an exclusivity provision for interchangeable products is implemented. These anticompetitive consequences are likely to result if the period can be extended, the period does not run immediately upon its award, or if a firm has the ability to delay triggering the running of the period through, for example, a patent settlement, acquisition, merger, or agreement.[303] In addition, each of these problems is likely to be present even if the exclusivity is awarded on a "first-to-file" rather than a "first-to-apply" basis.

For example, pioneer manufacturers and FOB applicants could settle patent litigation such that a payment is made to the first interchangeable FOB entrant to settle the patent dispute and defer its entry. This settlement could create a bottleneck that blocks subsequent interchangeable FOB from obtaining FDA approval because the first-approved product's exclusivity period has not run. This outcome results in significant harm to consumers who not only lose the benefit of the first interchangeable product's entry but also the second product's entry. Furthermore, in this circumstance, the rationale for the FOB exclusivity period is undermined by proof that the subsequent applicant did not need an additional incentive to perform all the steps necessary to enter the market, yet is blocked from the market by the first interchangeable product.

In theory, various regulatory fixes could require an interchangeable FOB manufacturer to forfeit its exclusivity period. These forfeiture events could include when: (a) it fails to trigger the running of the period by launching the interchangeable FOB product immediately following a final court decision in its favor on the patents at issue; (b) it has not been sued by the branded manufacturer, or (c) its patent suit is taking too long to resolve and a subsequent interchangeable applicant is approvable by the FDA.

The problem with these fixes is that each one blocks entry of a subsequent interchangeable product for a period of time and thereby denies consumers price competition and increased innovation. They also require the FDA to expend significant resources monitoring patent registrations and certifications, litigations, and marketplace activity that is outside its core missions and competencies.[304] Further, such a marketing exclusivity provision will inevitably generate lawsuits against the FDA regarding award, timing, scope and termination of the marketing exclusivity periods as has occurred regarding the 180-day provision of the Hatch-Waxman Amendments.[305]

For these reasons, an FOB exclusivity period is unlikely to benefit consumers either with an increase in the pace or scope of innovation or additional price competition. An FOB exclusivity period is likely to delay FOB competition in the case when a second interchangeable FOB applicant is ready to be approved, but cannot enter until the first-approved interchangeable product's exclusivity has expired.

Appendix A

Economists have developed a framework to calculate the time it takes for a branded biologic drug manufacturer to recover fully its investment to develop and commercialize a typical biologic drug. This framework is referred to as the "Nature model" because it first appeared in an article by Dr. Henry Grabowski in the journal *Nature Reviews: Drug Discovery*.[306] The original Nature model, along with subsequent suggested changes, has been used as the basis for an estimation of the optimal length of a branded exclusivity period.[307]

This appendix describes the Nature model and explains the methodological and conceptual weaknesses that render its results too imprecise and non-robust to inform discussions about the length of a branded exclusivity period. A model that balances the benefits of FOB competition with the costs of potentially forsaking marginal branded drug development projects would be more informative than the Nature model's approach.[308]

The appendix is organized as follows: Section I describes the original Nature model's data inputs and the operation of the model; Section II describes the comments about, suggested changes to, and subsequent sensitivity analysis performed on the original Nature model; Section III describes the current weaknesses with the model; and despite these weaknesses, Section IV presents one correction to the elasticity and internal consistency flaws of the model along with new results based on these corrections.

I. Description of the Nature Model

The Nature model calculates the break-even point for a branded manufacturer's biologic portfolio as the point at which the net present value of the cumulative cash flows of the portfolio equals zero. The stream of cash flows upon which this calculation is based has the following six components.

A. Total Pre-Approval Research and Development Costs per Approved Biologic Drug

The first input of the Nature model is an estimate of the pre-approval R&D costs per approved biologic drug based on work by DiMasi and Grabowski.[309] The original Nature model and subsequent calculations rely on estimates of the total R&D costs for a typical investigational drug, adjusted for the probability of FDA approval, to calculate an estimate of the total R&D costs for a FDA-approved branded biologic drug. The R&D cost estimates are based on the proprietary data for 17 biologic drugs. The cost estimates are the weighted average costs in each phase of development (*i.e.*, preclinical, Phase I, Phase II, and Phase III) across the 17 drugs, where the weights are the probabilities of entering each phase.[310] The estimated real (*i.e.*, in 2005 dollars) costs for each phase are: $59.88 million (preclinical), $32.28 million (phase I), $31.55 million (phase II), and $45.26 million (phase III). These costs would be spent over an average 13 year period prior to approval, so that the future value of these costs at the time of approval (using a discount rate of 11.5%) is $374.70 million.

Because every molecule developed is not approved (*e.g.*, clinical testing may show that it is not safe and/or effective), the total R&D estimate is adjusted for the probability of success. The R&D cost per investigational molecule is converted to an estimate of the R&D cost per *approved* molecule by dividing the $374.70 million by the estimated probability of success (30.2%). The overall estimate of the total pre-approval R&D costs at launch (using a discount rate of 11.5%) for a typical approved biologic drug is $1.24 billion. Using a discount rate of 12.5%, the estimate of pre-approval R&D costs is $1.33 billion.

B. Launch and Plant Transition Costs

The second input of the Nature model is an estimate of the costs of launching production of the new drug. The Nature model and subsequent calculations assumed that the branded manufacturer will spend $25 million over the two years prior to launch to convert existing manufacturing capacity to the production of the new drug.[311] It is also assumed that the branded manufacturer will incur additional "launch-related expenditures equal to 10% and 20% of first year's sales" in the two years prior to launch, respectively.[312] Using a discount rate of 11.5%, the future value of these costs at the time of launch is roughly $70 million. Therefore, if the discount rate is assumed to be 11.5%, the typical branded biologic firm is estimated to be "in the hole" roughly $1.31 billion per approved new drug at the time of launch ($1.24 billion plus $70 million based on a 11.5% discount rate).

C. Post-Approval R&D Costs

The third input of the Nature model is an estimate of the costs for post-approval R&D of new indications and formulations. The Nature model and subsequent calculations assumed that these costs are $24.5 million per year for the first 8 years after approval based on post-approval R&D expenses of traditional small-molecule drug companies.

D. Revenues

The fourth input of the Nature model is an estimate of the revenue stream used to recover the pre-approval R&D costs, launch and transition costs, and post-approval R&D costs. The Nature model's revenue estimates are based on revenues from a sample of 30 biotechnology drugs. The 30 drugs are ranked into quintiles and the mean amounts for the top four ranked quintiles are then used to calculate the average revenue profile for a typical branded biologic drug. The Nature model excluded the bottom quintile because these drugs "may not have representative R&D cost profiles."[313]

The timing of the revenues for the hypothetical portfolio is assumed to match that of the "average new drug introduction in the 1990s." After the maximum revenues are achieved in the tenth year after launch, revenues are assumed to decline by 3.5% per year due to "obsolescence and therapeutic class competition."[314] The revenue stream represents worldwide sales and is denominated in 2005 dollars.[315]

E. Contribution Margin

The fifth input of the Nature model is an estimate of the operating profit margin, or contribution margin, of the brand drug. After the brand drug is launched, its revenues cover its operating costs each year with the remaining operating profit contributing to the recoupment of the investment costs. The original Nature model assumes that the contribution margin for the biologic portfolio is -30% in the first year after launch, +20% in the second year after launch, and +50% thereafter. The steady-state 50% margin is used because it is "in line with the contribution margins realized by the eight largest biotechnology firms with multiple products on the market."[316]

F. Discount Rate/Cost of Capital

The final input of the Nature model is an estimate of the cost of capital for a biologic drug, *i.e.*, the rate of return required by investors to compensate for the risk involved in developing the brand drug. The original Nature model uses two rates to "capitalize forward" the R&D cost stream to the launch date and discount the profit stream back to the launch date: 11.5% and 12.5%. These rates are justified as "reflective of the equity cost of capital for larger publicly listed biotechnology firms with multiple products on the market in recent periods."[317] These rates are based on estimates of the real cost of capital over time for an unspecified sample of biotech firms calculated in previous research using the capital asset pricing model (12.5% in 1994, 12.0% in 2000, and 10.0% in 2004).[318] The original Nature model uses the average (11.5%) and maximum (12.5%) of these three rates in its calculations.

Table 1. Break-Even Analysis w/ Discount Rate = 11.5% (all amounts in millions)

Year	R&D	Revenue	Margin	Post-approval R&D	Launch/ Plant	Profit	NPV Profit	Cumulative NPV
0	-$1,243				-$70	-$1,313	-$1,313	-$1,313
1		$128	-0.3	-$24.5		-$63	-$60	-$1,372
2		$243	0.2	-$24.5		$24	$20	-$1,352
3		$328	0.5	-$24.5		$140	$106	-$1,245
4		$413	0.5	-$24.5		$182	$124	-$1,121
5		$506	0.5	-$24.5		$229	$140	-$981
6		$577	0.5	-$24.5		$264	$145	-$836
7		$648	0.5	-$24.5		$300	$148	-$688
8		$676	0.5	-$24.5		$314	$139	-$550
9		$713	0.5			$357	$141	-$409
10		$713	0.5			$357	$127	-$282
11		$688	0.5			$344	$110	-$172
12		$664	0.5			$332	$95	-$77
13		$641	0.5			$320	$82	$5
14		$618	0.5			$309	$71	$76
15		$597	0.5			$298	$62	$138
16		$576	0.5			$288	$53	$191
17		$556	0.5			$278	$46	$237

These six components are used to calculate the point at which R&D costs are recouped through post-approval cumulative profits as shown in Table 1 below (which assumes a discount rate of 11.5%). The typical biologic product starts out at launch $1.31 billion "in the hole." During each year after launch, it earns an operating profit (assumed to be negative in the first year) which is its contribution margin times its revenue (minus any post-approval R&D). To properly compare profits in different years, this profit stream is discounted back to the launch date. If 11.5% is used as the discount rate, the cumulative profit stream covers the initial R&D expenditures late in the 13th year after launch (*i.e.*, 12.9 years after launch).[319] If 12.5% is used as the discount rate, this break-even point occurs 16.2 years after launch.

Based on these two calculations, Grabowski concludes that the exclusivity period for a branded biologic should last 12 and 16 years after launch: "entry through abbreviated filings should be delayed until the representative NBE [New Biologic Entity] has had the opportunity to earn risk-adjusted break-even returns."[320]

II. Summary of Comments

Commenters raised several issues about the inputs, the operation of the model, and the inferences that can be drawn from the model. Alex Brill arrived at different results by varying some of the model's assumptions.[321] First, he suggested that 10% is a more accurate estimate of the cost of capital for biotech firms, rather than 11.5% or 12.5%.[322] Second, he posited that 60%, rather than 50%, is a more accurate estimate of the contribution margin for a large biotech firm.[323] Using these assumptions, he calculates a break-even point nine years after FDA approval of the branded biologic drug.

Brill also explained that the break-even point should not be used as a proxy for the optimal exclusivity period because a branded biologic product is likely to continue earning positive profits even after FOB entry.[324] If exclusivity is granted so that no FOBs can enter until the average branded manufacturer has recouped its R&D costs, then the branded manufacturer will earn "profits that exceed the required rate of return expected by investors."[325] If true, this could lead to consumer harm through the delay of FOB entry. As an illustration, Brill used the 10%/60% discount rate/margin assumptions, along with assumptions from the Congressional Budget Office about the market share and price declines that branded biologic drugs are likely to face with FOB entry.[326] This illustrations shows that a branded manufacturer would break-even ten years after approval even if FOB entry occurs in the eighth year after FDA approval of the branded drug.

Another commenter argued that the use of a portfolio approach in estimating the revenue stream may result in an exclusivity period that is too long and overprotects the branded biologic drugs that are most likely to face FOB competition. Although a branded manufacturer and its capital partners may diversify by investing in many investigational drugs (some of which will become very successful and profitable and others that will be approved, but have relatively small revenues), potential FOB entry is only a credible concern for the most successful of these drugs. The portfolio used in any break-even calculation to determine exclusivity periods should only include those drugs for which FOB entry is likely when the period of exclusivity expires. The original Nature model and subsequent calculations exclude only the bottom quintile of biologic drugs when constructing the portfolio. This commenter suggested that drugs with less than $250 million in sales are unlikely to face FOB competition, thus implying that the bottom two quintiles should be excluded from the break-even calculation, as the drugs in the second lowest quintile have peak sales of $100 million.[327]

Following the roundtable and in response to Brill's critique, Grabowski questioned Brill's assumption about total market revenues when FOBs enter the market; cited additional research suggesting that the true cost of capital may be higher than originally presented in the Nature model; suggested that the true contribution margin may be lower than originally presented; and provided additional analysis showing how relaxing other assumptions in the model would lead to longer break-even times.[328]

Brill also provided post-roundtable comments that included additional alternative interpretations of CBO's assumptions regarding branded manufacturer market share declines following FOB entry.[329] In addition to his original calculation (which assumes total market revenues do not change), he presented additional break-even calculations that assume a perfectly inelastic demand. One of these assumes a steady-state price decline of 40% as before and the other assumes a steady-state price decline of 20%. A final calculation assumes no price decline and simply a loss of market share to the FOB entrant. As in his original analysis, all of these calculations assume seven years of branded exclusivity, a contribution margin of 60%, and a cost of capital of 10%. With these assumptions, he finds that the branded manufacturer breaks-even nine to 14 years after launch.

Finally, one commenter surmised that the Nature model was "fraught with peril." He suggested that: "Aside from a possibly nonrepresentative sample, the exercise involves numerous assumptions about the cost of capital, profit margins, and prices after the first follow-on enters the market. Reasonable changes to these assumptions can easily affect the results by 30–40 percent."[330]

III. Problems with the Break-Even Model - Analysis

The problems with the Nature model fall into three types: (1) problems with the "inputs" to the model (*i.e.*, problems with the underlying components); (2) problems with the incorporation of FOB entry into the model after a period of exclusivity; and (3) problems with the interpretation and use of the results.

A. Problems with the Input Assumptions

The first input to the Nature model is the estimate of pre-approval R&D costs for a representative biologic drug. The problem with this estimate is that it is based on a sample of only 17 drugs. No variance information is presented for the sample.[331] Unless the R&D costs within each clinical phase are essentially identical across the drugs, it is likely that the confidence interval around the R&D cost estimate is large and, thus, the R&D cost estimate is less likely to be accurate. Further, 13 of the 17 drugs were developed by one firm and the sample is restricted to therapeutic recombinant proteins and monoclonal antibodies, so it is possible that the sample is non-random and not representative of biologic drugs overall.

Another important input into the model is the revenue stream of the representative biologic drug. There are two potential problems with the revenue stream used in the Nature model.[332] First, the revenue stream includes sales from post-approval indications and formulations in addition to the original indication/formulation.[333] The revenue stream associated with the original indication/formulation is not provided, so one cannot calculate the break-even point of the original indication/formulation with the data in the model.

Second, the revenue stream assumes "therapeutic class competition."[334] The time frame of the therapeutic class competition is not stated in the original Nature model, but Grabowski's post-conference comment suggests that the therapeutic class competition is assumed to begin in the tenth year after approval of the brand drug.[335] Thus, the original Nature model implies that a "first-in-class" branded drug will recoup its R&D costs even if therapeutic class competition occurs. And because entry of FOBs is likely to have the same market effect as entry by branded competitors, this assumption leads to the conclusion that the branded exclusivity period for the first-in-class branded drug should be less than 12 to 16 years.

In addition, like the R&D cost estimates, the revenue estimates are based on a small sample. The model relies on 24 biologic drugs to estimate the revenue stream. These are the drugs in the top four quintiles of the distribution and the spread in average peak revenues between the top and second-to-bottom quintiles ($2 billion to $100 million) suggests a large variance in this distribution. Like the R&D cost estimates, it is likely that the confidence interval around the estimated revenue stream is large and, thus, the revenue estimates are less likely to be accurate. In addition, it is unknown whether the 24 drugs are a random sample of biologics. If not, or if they substantially overlap with the 17 drugs used to estimate the R&D costs, the revenue estimates may be biased like the R&D cost estimates.

Furthermore, an implicit assumption of the Nature model is that there is no correlation between R&D costs and revenues, so that an average R&D cost stream and an average revenue stream can be used to make inferences about average profitability. However, R&D costs and revenues may be positively or negatively correlated, making the variance of the profit estimates smaller or larger, respectively, than suggested by independent samples. Since the samples used to estimate R&D costs and revenues are not disclosed, it is impossible to determine if this ameliorates or exacerbates the measurement error.

Another important component in the model is the assumed cost of capital. Despite the disagreement over the appropriate cost of capital for a biologic firm, the model assumes a constant cost of capital throughout the entire product life cycle. Investments in biologic R&D during the early stages of research (*e.g.*, preclinical R&D) might have a higher cost of capital reflecting their relative risk, while investments during the later stages (*e.g.*, phase III and post-approval) might have a lower cost of capital reflecting the relative certainty of the return.[336] As a result, this could substantially change the total capitalized amount to be recouped.

Apart from the estimation errors associated with the R&D cost and revenue inputs and the cost of capital assumptions, it is unclear whether the use of sample means or a portfolio approach is appropriate. The samples of drug projects used to create the R&D cost and revenue estimates are biologic development projects that were actually pursued. At least initially, all of these projects were perceived as potentially profitable ventures or else they would not have been funded. Introducing a FOB pathway might make some of these formerly profitable projects unprofitable. If the R&D cost and revenue figures are independent, the use of sample means in the Nature model imply that the exclusivity period should be set so that development of projects that are above the profit mean continue to be pursued while those below the profit mean are abandoned. If the R&D cost and revenue estimates are not independent (*e.g.*, more expensive projects are associated with larger expected sales), it is not clear which projects would be abandoned with a FOB pathway determined using the R&D cost and revenue mean estimates. A better approach to determining the optimal length of a branded exclusivity period would balance the benefits of FOB competition with the costs of potentially forsaking marginal branded development projects.

B. Problems Incorporating FOB Entry into the Break-Even Model

The versions of the model that explicitly incorporate FOB entry contain two questionable assumptions that may bias the results. First, in most of the calculations that incorporate FOB entry, it is assumed that the price of the branded drug will gradually decline after FOB entry so that it is 40% below its pre-FOB price in four years. This assumption corresponds roughly to the CBO's assumption that the FOB price will be 40% less than the pre-FOB branded price four years after entry. Of course, the latter does not necessarily imply the former, as it is theoretically possible for the FOB's price to be 40% less than the branded drug's pre-FOB price even if the branded drug's price falls by more or less than 40%.

The assumption that the branded drug price will match the FOB price represents the least profitable scenario for the branded manufacturer: the scenario in which the branded drug and the FOB are perfect substitutes.[337] Still, the 40% branded price decrease assumption is ad hoc and is not necessarily consistent with the CBO assumption that the branded drug's market share will eventually decline by 35%. In the calculations of Grabowski and Brill, competition between the branded manufacturer and the FOB firm following FOB entry is not modeled explicitly. It is reasonable to expect that the branded drug's price decrease and market share decrease are interrelated and are jointly determined. In other words, the assumption that the branded drug's share declines by 35% in large part determines what price decreases are possible. The analysis below provides one example of how the Nature model can be corrected to account for this relationship.

The second questionable assumption concerns overall revenues after FOB entry. In some variations of the model, it is assumed that overall market revenues stay the same after FOB entry. In other words, it is assumed that the branded manufacturer faces a demand for its drug that is unitary elastic so that price decreases and the resulting increases in the quantity demanded exactly offset producing no change in overall market revenue. For profit-maximizing firms, this is impossible as profit-maximizing firms (with positive marginal costs) always price in the elastic portion of the demand they face.

In other variations of the model, it is assumed that overall revenues decline by the same fraction as the price which is equivalent to assuming that the demand for the branded drug is perfectly inelastic (*i.e.*, a price decline results in no change in the quantity produced and sold). A profit-maximizing firm facing a perfectly inelastic demand will increase its price to infinity, as it can sell the same amount at a higher and higher price. This result also is impossible. Branded manufacturers set prices in the elastic portion of the demands they face for the drugs they produce, as do all profit-maximizing firms.[338]

This latter assumption is problematic because it assumes away the primary benefit of establishing an abbreviated pathway for follow-on biologics, namely, to reduce the price of biologic drugs so more people can have access to them. Furthermore, this assumption of inelastic demand directly contradicts the contribution margins used in the model. The Lerner Index dictates that for any profit-maximizing firm, its profit margin over marginal cost will equal the inverse of (the absolute value of) the price elasticity of demand for the demand the firm faces for its product. This condition holds for all profit-maximizing firms, not just monopolists, as it is derived from the first-order necessary condition for profit-maximization. Thus, any profit-maximizing firm that has positive marginal costs and a finite, positive margin over marginal cost must be facing an elastic demand (at least locally around the profit-maximizing price). The contribution margins used in the model are not necessarily equal to the margins over marginal cost used in the Lerner Index (*e.g.*, they may include some overhead costs). However, the assumption of finite contribution margins necessarily implies finite margins over marginal cost and, thus, demand that is elastic, not perfectly inelastic. Below, we correct the model's calculations using the Lerner Index with the contribution margin serving as a proxy for the margin over marginal cost.

C. Problems in the Interpretation and Use of the Results

Apart from the problems with the underlying assumptions of the model, the results of the model are prone to misinterpretation. First, the inclusion of post-approval R&D costs and revenues in the break-even analysis makes it easy to misinterpret the results if one is using the analysis to determine the extent of the exclusivity period for branded biologics. If a fixed exclusivity period is set to recoup the costs of pre-approval and post-approval R&D, then the exclusivity period provides no marginal incentive to the branded firm to conduct post-approval R&D.

Theoretically, the preceding issue could be resolved if one were able to separate the revenues from post-approval indications and formulations from the revenues for the original indication and formulation. However, even if one were able to correct this problem and all of the previous problems, a more fundamental problem with the general model remains: small changes in the input assumptions yield large swings in the resulting break-even period.

This last point can be illustrated by considering Grabowski's post-conference comment that the actual cost of capital for biotech firms may be as high as 13-15%. Recall that the Nature model originally found that a representative biologic portfolio would break even in the 13th year if the cost of capital was 11.5% and would break even in the 17th year if the cost of capital was 12.5%. Using identical assumptions to the original Nature model, but increasing the cost of capital to 13.25% produces a break-even point in the 23rd year. If the cost of capital is greater than or equal to 13.7%, and all of the other assumptions in the original model are retained (including no FOB entry at any point), the representative biologic portfolio *never* breaks even. The fact that the representative biologic portfolio *never* breaks even when using a cost of capital greater than or equal to 13.7%, even though Grabowski's post-roundtable comments suggest 14-15% is a plausible cost of capital for biotech firms, casts doubt on the accuracy and reliability of the model.

IV. Correcting Problems in the Nature Model Does Not Improve its Usefulness

As discussed above, the break-even calculations of the Nature model suffer from many problems. Some of these problems can be corrected. In particular, the assumptions of unitary elastic and perfectly inelastic demand can be discarded to make the model's elasticity assumptions consistent with its contribution margin assumptions. Second, the model's consistency problems in the post-FOB world can be corrected by applying a reasonable and flexible competition model. The corrections described below are not exhaustive and simply represent one way these problems can be addressed. In fact, the assumption of Cournot FOB/branded competition is likely wrong, but is a reasonable approach that is consistent with the assumption made by both Grabowski and Brill that the FOB and branded drugs will have the same price after FOB entry. However, the corrections illustrate that the elasticity and consistency problems are not innocuous, but instead have a significant impact on the results. These corrections do not address the more fundamental problems of imprecision and non-robustness. As such, even with these corrections, we find the break-even framework uninformative in the debate about proper exclusivity periods for branded biologic drugs.

First, regarding the elasticity assumption, assume that the contribution margin of the branded manufacturer is equal to the branded manufacturer's margin over marginal cost (p-mc)/p. These two margins are probably not equal, as the former includes some overhead costs, but the contribution margin is the best proxy for the margin over marginal cost that is readily available. From the Lerner Index, the price elasticity of demand for the branded manufacturer's drug is (-1 times) the inverse of the contribution margin. In other words, the assumption of a contribution margin of 50% implies an elasticity of -2 and a contribution margin of 60% implies an elasticity of -5/3. Let the subscript 1 denote the period before FOB entry and let the subscript 2 denote the period after FOB entry. Following Grabowski and Brill, we assume that the branded drug's price following FOB entry is the same as the FOB's price.[339] Let α be such that $p_2 = \alpha p_1$. For example, a steady-state price decline of 40% following FOB entry implies that $\alpha = 0.6$. The price elasticity of demand (ε) is the ratio of the percentage change in quantity with a corresponding percentage change in price. Using an approximation of the elasticity and normalizing $q_1 = 1$, this implies:

$$\varepsilon = \frac{\dfrac{q_2 - q_1}{q_1}}{\dfrac{p_2 - p_1}{p_1}} \Rightarrow q_2 = \varepsilon(\alpha - 1) + 1 \tag{1}$$

This implies that the total market (*i.e.*, branded + FOB) revenue after FOB entry is:

$$R_2 = p_2 q_2 = \alpha[\varepsilon(\alpha - 1) + 1]R_1 \tag{2}$$

The profit-maximizing branded manufacturer's marginal cost will equal its marginal revenue before FOB entry:

$$mc = mr = p_1\left(1 + \frac{1}{\varepsilon}\right) \tag{3}$$

Therefore, assuming the branded manufacturer's marginal cost does not change after FOB entry, the branded manufacturers's post-FOB margin will be:

$$\frac{p_2 - mc}{p_2} = \frac{\alpha - \left(1 + \dfrac{1}{\varepsilon}\right)}{\alpha} \tag{4}$$

Second, a model of the competition between the branded firm and the FOB entrants is needed to characterize prices and market shares that are consistent with each other. There are a number of models that could be used, but the Cournot model seems most appropriate in the current context for the following reasons:

- All versions of the Nature model that incorporate FOB entry assume that the branded firm and the FOB will have the same price after FOB entry. This assumption is likely incorrect as it is likely that the brand and FOBs will not be perfect substitutes and the brand may continue to price higher than the FOBs. However, this assumption is used because it represents the least profitable scenario for the branded manufacturer.
- Most analysts expect that the price of biologics will decline as more FOBs enter, which is a characteristic of the Cournot model.[340]
- The Cournot model is flexible and requires limited assumptions to implement. It is robust across most types of demand and cost functions. Other potential models (e.g., differentiated Bertrand) require more assumptions to implement. However, if feasible to implement, a monopolistic competition model would be more accurate, as it would capture the differentiation between the brand drug and the FOB that is likely to characterize competition after FOB entry.

The Cournot analogue to the Lerner Index dictates that a firm's margin over marginal cost will equal (the absolute value of) the inverse of the price elasticity of demand times the firm's market share. Therefore, after FOB entry, the branded manufacturer's margin is:

$$\frac{p_2 - mc}{p_2} = -\frac{s}{\varepsilon} \tag{5}$$

where s is the branded manufacturer's market share. Substituting the branded manufacturer's marginal cost as calculated in (3) above, we can solve for the branded manufacturer's post-FOB price as a function of the branded pre-FOB price:

$$\frac{p_2 - \left[p_1\left(1 + \frac{1}{\varepsilon}\right)\right]}{p_2} = -\frac{s}{\varepsilon} \Rightarrow p_2 = \frac{(\varepsilon + 1)}{(\varepsilon + s)} p_1 \tag{6}$$

In other words, if the competition between the branded firm and the FOB firms is consistent with the Cournot model, assumptions about the brand's pre-FOB margin (which determines ε) and the brand's post-FOB market share uniquely determine the brand's price decrease after FOB entry. The CBO's assumption of a steady-state market share decline of 35% and the assumption of a 50% margin imply a 26% branded manufacturer price decrease. If a 60% margin is used instead, the branded manufacturer's price decrease is roughly 34%. Using these values in the break-even calculations produces the following break-even times:

Table 2. Year After Launch in Which Branded manufacturer Breaks-Even (Data Exclusivity = 7 Years)

Margin/Cost of Capital	Lerner (ε = -1/margin); 35% share decline; Cournot competition	Unitary elasticity; 35% share decline; 40% price decline	Perfect inelasticity; 35% share decline; 40% share decline
50%/10%	14	Never	Never
50%/11.5%	34	Never	Never
50%/12.5%	Never	Never	Never
60%/10%	10	11	14
60%/11.5%	14	17	Never
60%/12.5%	24	Never	Never

When the elasticity assumption is corrected and the price and share declines are made internally consistent, the results are much different than in previous versions of the Nature model. For example, Grabowski concludes that "notably, with an exclusivity period of 7 years, the *only* combination of assumptions that yields a breakeven point of less than 50 years is the one used by Brill."[341] On the contrary, with an exclusivity period of seven years, there is only one set of assumptions (of those most commonly used) that does not result in the branded manufacturer breaking-even.

However, even when the elasticity assumption is corrected and the price and share declines are made internally consistent, the break-even period varies from 10 years to infinity. Small changes to the margin and cost of capital assumptions cause large swings in the results. Under the original assumptions of a 50% margin and a 11.5% cost of capital, the brand biologic breaks even after 34 years if the exclusivity period is seven years. Increase the cost of capital assumption to 13.7% and the brand would *never* recoup its investments, even if exclusivity were perpetual and FOB's *never* entered. Note also that these large swings in the results occur even when one assumes the underlying cost and revenue estimates are measured without error. If one were to incorporate the large estimation errors that likely exist because of the small samples on which the estimates are based, the range of plausible results would only expand. A model that produces such vastly different results with small and reasonable changes in the underlying assumptions is unreliable as a basis for policy.

Appendix B

I. FDA's Drug Approval Processes

The Food and Drug Administration ("FDA") approves prescription drug medicines for marketing in the United States through two separate and distinct product approval pathways, depending on the drug's method of manufacture. The first pathway applies to small molecule drugs and the second pathway applies to biologic drugs.

Small molecule drugs are manufactured by chemical synthesis. The FDA's Center for Drug Evaluation and Research ("CDER") approves small molecule drugs pursuant to the Federal Food, Drug, and Cosmetic Act ("FD&C Act").[342] To obtain FDA approval, the small molecule drug manufacturer, or company sponsor, must complete the requirements of a full New Drug Application ("NDA"), including a showing of medical benefit over patient risk.[343]

Biologic products are derived from living matter (*e.g.*, purified from blood) or manufactured in living cells (*e.g.*, yeast, e.coli, or mammalian cells) using recombinant DNA biotechnologies.[344] The FDA's Center for Biologics Evaluation and Research ("CBER") approves biologic drugs for marketing pursuant to the Public Health Safety (PHS Act").[345] To obtain FDA approval, the company sponsor must complete the requirements of a Biologics License Application (BLA). The FD&C Act also regulates biologic products because most biologic products also meet the FD&C Act's definition of "drugs".[346]

The small molecule and biologic pathways have one main difference: the FDA may approve generic small molecule drugs using an abbreviated pathway, but no abbreviated process exists for follow-on biologic drugs.

A. New Drug Approvals

Although new small-molecule drug applicants file an NDA and new biologic drug applicants file a BLA, the development and regulatory approval process is similar for both categories.[347] For example, both small molecule and biologic drug applicants must establish medical benefit over patient risk.[348] The applicant, or company sponsor, also must prove the product is safe and effective. To do so, the applicant submits an NDA or BLA that contains the following information: (a) preclinical analytical tests, preclinical studies, and formulation studies; (c) an Investigational New Drug Application ("IND") to initiate human clinical testing; (d) adequate and well-controlled human clinical trials to establish the safety and efficacy of the drug for its intended use; (e) approval and validation of manufacturing facilities used in production of the pharmaceutical product; (f) drug manufacture and analytical methods; and (g) proposed product packaging and labeling.[349]

The preclinical phase of any new drug development typically begins with assays and large scale screening of compounds against targets of interest. Once a lead compound is isolated, preclinical safety trials are conducted, as well as trials in predictive animal models. This preclinical phase typically takes one to five years.[350] After preclinical tests are completed, a drug sponsor submits the results in an IND to the FDA for approval before human clinical trials begin.

Human clinical trials typically consist of three phases. In Phase I clinical trials, a small group of healthy human patients are given the drug to determine if the drug is safe in humans.[351] In Phase II clinical trials, a small sample of the intended patient population are given doses of the drug to provide a preliminary assessment of the efficacy of the drug for a specific clinical indication, find dose tolerance, and determine the optimal dose range. Safety data also is collected in Phase II as it is in all phases of drug testing.[352] Phase III studies are initiated if Phase I and Phase II studies indicate the drug is safe, and has some efficacy in the targeted patient population. Phase III clinical trials are designed to gather sufficient data in a broad target population in order to establish safety and efficacy for a particular indication.[353]

The time needed to conduct these trials varies based on factors such as indication, availability of reliable biomarkers to measure efficacy, patient size, and ease of patient accrual. Phase I trials generally take one to two years. Phase II trials, including a full dose ranging study take two to three years. Phase III trials are the longest, taking approximately three to five years. Time variability, however, is significant as efficacy burdens vary. For example, it takes less time to collect the data using an accepted biomarker, such as blood cell levels, to measure efficacy of a treatment than it does for to collect data measuring disease free progression, mortality and morbidity data. Drug products also are subject to marketing exclusivities, described in more detail below.

B. Abbreviated Drug Approvals for Follow-on and Generic Products

Prior to 1984, no process existed for abbreviated approval of generic small-molecule drugs. Generic versions of drugs approved after 1962 could only be approved pursuant to either a full New Drug Application or a "paper NDA" application under Section 505(b)(2) of the FD&C Act.[354] As a result, few companies developed generic drugs because of the high cost to perform the required clinical trials.[355]

In 1984, Congress enacted the Hatch-Waxman Amendments to the FD&C Act ("Hatch-Waxman") which established an abbreviated regulatory pathway to approve generic drug versions of drugs approved under that act. Hatch-Waxman provided the FDA with discretionary authority to not require generic applicants duplicate the safety and efficacy trials of the reference drug. Rather the Hatch-Waxman Act authorized the FDA to rely on its prior findings of safety and efficacy of previously approved drug products when the agency later reviewed the generic drug's application.[356] The Hatch-Waxman Act reflected Congress' attempt to balance the need to encourage innovation with the desire to speed the availability of lower cost alternatives to approved drugs."[357]

1. The Section 505(b)(2) "Paper NDA" Pathway

The 505(b)(2) or "paper NDA" pathway is a partially-abbreviated pathway for drugs that are similar to, but not copies of, a reference small-molecule drug. This pathway pre-existed the Hatch-Waxman Act. A 505(b)(2) applicant relies on one or more safety or efficacy investigations that were not conducted by the 505(b)(2) applicant, and for which the 505(b)(2) applicant has not obtained a right of reference, *e.g.*, reliance on results in the published literature. This pathway is especially useful for new dosage forms, strengths, rates of administrations, dosing regimens and new indications.

The 505(b)(2) pathway permits the FDA to rely "to the greatest extent possible on what is already known about a drug" so as to avoid requiring drug sponsors to conduct and submit studies that "are not scientifically necessary." FDA has stated that many of the drugs approved via the 505(b)(2) route would never have reached the market, or would have been significantly delayed, without this pathway.[358] Indeed, five significant FDA-identified harms could occur without the 505(b)(2) pathway: (1) diversion of industry resources that could otherwise be used to undertake innovative research; (2) increased drug costs; (3) strain on FDA review resources; (4) slowing of the process for drug approval with no corresponding benefit to the public health; and (5) significant ethical concerns raised by requiring duplicative studies that subject human beings and animals to medically and scientifically unjustified testing.[359]

2. The 505(j) ANDA Pathway

As discussed above, in 1984, Congress created an abbreviated pathway for approval of generic small-molecule drugs, this also is known as the 505(j) ANDA Pathway. Hatch-Waxman was designed to facilitate competition from lower-priced generic drugs, while maintaining incentives for pharmaceutical companies to invest in developing new drugs.[360] It also gave FDA discretionary authority to review an abbreviated new-drug application

("ANDA") for generic small molecule drugs.[361] This "reflected Congress' attempt to balance the need to encourage innovation with the desire to speed the availability of lower cost alternatives to approved drugs."[362]

Under the Hatch-Waxman Act, a generic drug applicant only is required to show that its product includes the same active ingredient(s) and is bioequivalent to a reference drug, but it does not need to replicate the clinical trials and other testing of the reference product.[363] This process typically involves bioequivalency trials in healthy human volunteers, showing that a generic drug has the same levels of the same active pharmaceutical ingredient as the reference branded product. Because reference and ANDA drugs must have the same or similar API, dosage forms, strength, route of administration, labeling, quality, performance and intended use duplicate clinical trials are unnecessary. State substitution laws allow for the substitution of a bioequivalent generic product for the branded reference drug at the retail pharmacy without the doctor's involvement.

3. Patent Restoration and Patent Listings for New Drug Products

Before Hatch-Waxman, 505(b)(2) applicants could not begin preclinical or clinical trials until after patents expired on the relevant branded product without risking infringement of the branded product's patents. The risk of patent infringement coupled with the FDA generic approval process, in effect, extended the term of the branded company's patent protection and delayed market entry by follow-on applicants' versions of branded pharmaceutical drug products.[364] Hatch-Waxman limited the applicant's infringement liability so that it could begin the research, development, and manufacture of a drug product intended for FDA approval without infringing the branded product's patents.

Before 1984, branded pharmaceutical companies asserted that the effective terms of the patents covering their drugs were shortened due to the delays in the FDA approval process. To maintain incentives for branded drug product innovation in the face of generic competition, Congress included in the Hatch-Waxman Amendments patent restoration provisions that apply to drugs approved under both the FD&C Act and the PHS Act.[365] The extension period is calculated on the basis of length of time required to study and gain approval of the patented product. A maximum of five years can be restored to the patent. In all cases, the total patent life for the product with the patent extension cannot exceed 14 years from the product's approval date, or in other words, 14 years of potential marketing time. If the patent life of the product after approval has 14 or more years, the product would not be eligible for patent extension.[366]

Additionally, Hatch-Waxman provided operational provisions to encourage simultaneous running of the patent resolution process with any regulatory approval process, including marketing exclusivity periods.[367] To accomplish this, Hatch-Waxman amended the FDA's new drug approval process to require that the reference branded company list all of the reference drug's patents, and patent extensions.[368] Once these patents are listed in the Orange Book it is incumbent upon the follow-on applicants to certify how these patents relate to its drug product.[369]

If an ANDA or 505(b)(2) NDA applicant certifies that a referenced drug patent information has not been filed in the Orange Book, or that such patent has expired, then the FDA may approve the application immediately, provided other requirements are met.[370] If the applicant certifies that it will not launch its product until after the referenced product's patents expire, the FDA may approve the application effective on the date the patent expires.[371]

However, if an applicant makes a certification under Paragraph IV, Hatch-Waxman requires that the applicant to also provide notice to both the patent holder and the NDA filer.[372] Once the ANDA filer has provided such notice, a patent holder (usually the referenced branded company) must bring an infringement suit within 45 days to trigger the 30-month stay of FDA approval of the application. Hatch-Waxman provides a 30-month stay of FDA approval with a detailed statement of the factual and legal basis for the applicant's assertion that the patent is invalid or not infringed. If suit is not filed by that date, the FDA may approve the application. If patent infringement litigation is initiated by the branded product company within the 45-day period, then the FDA approval of the application is stayed until the earliest of: (1) the date the patent(s) expire; (2) a final determination of non-infringement or patent invalidity by a court in the patent litigation; or (3) the expiration of the 30 months from receipt of notice of the paragraph IV certification.[373]

4. Marketing Exclusivities

Under the Hatch-Waxman Act, FDA enforces five types of exclusivity: (1) new chemical entity – five years; (2) new clinical investigation – three years; (3) orphan drug – seven years; (4) pediatric – six months; and (5) ANDA patent challenge exclusivity – 180 days. The first four apply only to New Drug Application ("NDA") filers. The fifth applies only to Abbreviated New Drug Application ("ANDA"), *i.e.* generic, filers. The Orange Book lists all exclusivities granted to each approved-drug product.

New chemical entity (NCE) exclusivity provides five years of exclusivity, from the date of approval of the first NDA, for new drug applications containing new chemical entities never previously approved by FDA, either alone, or in combination.[374] An NCE is a drug that contains "no active moiety previously approved by the FDA."[375] No ANDA or 505(b)(2) application may be submitted during the five-year NCE period, except that such applications may be submitted after four years if they contain a certification of patent invalidity or non-infringement (*i.e.*, paragraph IV certifications).

NCE exclusivity is the only exclusivity that bars the FDA from even accepting applications for review (as opposed to allowing the submission and review of such applications and simply delaying FDA approval). The five-year exclusivity period does not bar the FDA from accepting another full competitor NDA if the sponsor of the second application has done all of the work itself. As a practical matter, NCE exclusivity delays competition for more than five (or four) years because, once the application has been submitted, it typically takes the FDA at least an additional year to review and approve the ANDA.

New clinical investigation (NCI) exclusivity grants three years of exclusivity for certain changes to a drug product.[376] It prohibits FDA from approving an application for the same product for three years.[377] This exclusivity begins at the approval of the product, and is limited to the changes in the product supported by the new clinical studies. To obtain NCI exclusivity, the application or supplement must contain reports of new clinical investigations conducted by the sponsor. Several requirements apply, including that the study be clinical (*i.e.*, in humans, not animals), that it be new (and generally not used for another drug approval purpose), and that it be essential to approval (*i.e.*, not merely interesting and useful).

Seven years of exclusivity also is available for Orphan drugs (*i.e.* drugs that treat a patient population with a target population less than 200,000).[378] The Orphan Drug Act of 1983 established an exclusivity period designed to provide an incentive to pharmaceutical manufacturers to develop drugs to treat rare diseases or conditions affecting relatively small numbers of persons. An orphan drug is defined as one treating a disease or condition which affects less than 200,000 persons in the United States or affects more than 200,000 persons but for which there is "no reasonable expectation that the cost of developing and making available in the United States a drug for such disease or condition will be recovered from sales in the United States of such drug."

Obtaining orphan drug status is a two-step process. First, the applicant must apply for, and receive, orphan-drug designation from the Office of Orphan Products Development at the FDA. Orphan designation qualifies the sponsor of the product for the tax credit and marketing incentives of the Act in exchange for developing the drug for a rare disease or condition. Second, like any other new drug, the orphan-designated drug must submit its full NDA for safety and efficacy review.

If the NDA is approved for the indication for which the orphan designation was granted, the developer of an orphan product receives seven years of market exclusivity following the approval of the product by the FDA. Orphan drug exclusivity protects the drug for the approved orphan indication against all other competitors. Unlike other exclusivities, orphan exclusivity protects the orphan drug even from a second full NDA for the same indication submitted by another applicant. Exclusivity applies only to the indication for which the drug has been designated and approved, however, so that a second application for the same drug for a different use could be approved by the FDA.

Any small molecule or biologic drug product can also obtain an additional six months of marketing exclusivity for demonstrating the safety, dosing and efficacy of the product in children.[379] Congress provided for a six-month pediatric exclusivity period in response to a perceived need for an incentive to encourage companies to complete and submit studies on the pediatric uses of drugs. Pediatric exclusivity attaches to all the applicant's formulations, dosage forms, and indications for products with existing marketing exclusivity or patent life that contains the same active moiety.

This is a broad grant because it attaches not only to the specific product that was studied in the pediatric population, but to all drug products (formulations, dosage forms, and indications) with the same active moiety. To balance this broad grant of exclusivity, the FDA requires pediatric studies of all drugs that contain the active moiety. This does not mean that a sponsor must show that the drug is safe or effective in the pediatric population to obtain pediatric exclusivity. Instead, the goal simply is to develop a maximum amount of pediatric information as a result of the grant of exclusivity. Pediatric exclusivity may therefore be granted upon acceptance of the pediatric study reports.

Pediatric exclusivity is unique because it attaches to the end of all existing marketing exclusivity and patent periods. This distinguishes it from other types of exclusivity and patent periods, which run concurrently. For example, if a drug sponsor has five-year NCE exclusivity (which is valuable because it bars competitors from even submitting applications to the FDA); the six-month pediatric exclusivity will provide six additional months of NCE exclusivity. If the drug sponsor has three years of new clinical investigation exclusivity, which bars the FDA from approving a competing application, the six-month pediatric exclusivity will provide six additional months of the same protection. If the drug sponsor has a patent, FDA-enforced exclusivity will be added at the end of the patent term.

The Hatch-Waxman Act grants 180 days of marketing exclusivity to certain generic drug applications. The statute provides an incentive of 180 days of market exclusivity to the “first” generic applicant who challenges a listed patent by filing a paragraph IV certification and therefore runs the risk of having to defend a patent infringement suit. As a practical matter, if multiple ANDA filers file paragraph IV certifications on the same day and all are found acceptable for filing, multiple applicants may share the 180-day exclusivity. The statute provides that the first application to file a substantially complete ANDA containing a paragraph IV certification to a listed patent will be eligible for a 180-day period of exclusivity beginning either from the date it begins commercial marketing of the generic drug product, or from the date of a court decision finding the patent invalid, unenforceable or not infringed, whichever is first. These two events – first commercial marketing and a court decision favorable to the generic – are often called "triggering" events, because under the statute they can trigger the beginning of the 180-day exclusivity period. Approval of the ANDA alone has no effect on triggering the 180-day patent exclusivity period.

If there is no court decision, and the first applicant does not begin commercial marketing of the generic drug, there may be prolonged or indefinite delays in the beginning of the first applicant's 180-day exclusivity period. Until an eligible ANDA applicant's 180-day exclusivity period has expired, the FDA cannot approve subsequently submitted ANDAs for the same drug, even if the later ANDAs are otherwise ready for approval and the sponsors are wiling to immediately begin marketing. Therefore, as a practical matter, an ANDA applicant who is eligible for exclusivity is often in the position to delay all generic competition for the branded drug.

In 2003, the Medicare Modernization Act amended the Hatch Waxman Act, to provide an 180-day market exclusivity period to the first generic company that seeks FDA approval to market at product prior to the expiration of certain patents relating the branded drug product. No other generic manufacturer may obtain FDA approval to market its product until the first generic applicant has sold its product for 180 days, unless the later generic applicant wins a patent challenge against the branded company.[380]

The Hatch-Waxman Act also provides marketing exclusivity incentives to any NDA holder based on the level of innovation represented by the drug product. Any "new chemical entity" receives five years of marketing exclusivity. During this five-year period, the FDA may not review any 505(b)(2) or 505(j) applications that reference this new chemical entity. However, if an ANDA files a paragraph II certification against this NCE, exclusivity is limited to four years.[381] The FDA may not approve this application until after seven and one half years or patent litigation is resolved.[382]

Appendix C. Public Comments FTC Roundtable on Competiton Issues Involving follow-On Biologic Drugs November 21, 2008

NAME	TITLE OF COMMENT
American Association of Retired Persons (David P. Sloane)	12/22/2008
American Enterprise Institute (John E. Calfee)	12/10/2008 • *When Patents Are Not Enough: Data Exclusivity for Follow-On Biologics*
Amgen, Inc. (David W. Beier)	9/30/2008
Barr Pharmaceuticals, Inc. (Bruce L. Downey)	9/30/2008
Barr Pharmaceuticals, Inc. (Bruce L. Downey)	12/19/2008
Bayer HealthCare LLC (Sandra S. Oliver)	10/2/2008
Bernstein Research (Ronny Gal)	12/9/2008 • *Eight Thoughts on Biosimilars*
Biotechnology Industry Organization (John M. Taylor, III)	9/30/2008
Biotechnology Industry Organization (John M. Taylor, III)	12/22/2008
Biotechnology Industry Organization (Sandra J. P. Dennis)	2/24/2009
Biotechnology Industry Organization (Sandra J. P. Dennis)	2/24/2009 • *Unpatentable Drugs and the Standards of Patentability, Benjamin N. Roin, (May 1, 2008)*
Biotechnology Industry Organization (Sandra J. P Dennis)	2/24/2009 • *Biosimilars, Data Exclusivity, and the Incentives for Innovation: A Critique of Kotlikoff's White Paper, Henry C. Grabowski and Joseph DiMasi,* (*February 2009)*
Biotechnology Industry Organization (Sandra J. P. Dennis)	2/24/2009 • *Data Exclusivity Period Length and Federal Government Savings from Enactment of the Biologics Price Competition and Innovation Act of 2007, Joseph Golec, John A. Vernon, and Ted Buckley,* (*January 28, 2009)*
Coalition for a Competitive Pharmaceutical Market (Annette Guarisco)	9/30/2008

(Continued)

NAME	TITLE OF COMMENT
CVS Caremark Corporation (David Golding)	12/22/2008
Duke University (Henry C. Grabowski)	12/22/2008
Duke University (Henry C. Grabowski)	12/23/2008 • *Data Exclusivity Periods for Biologics: Updating Prior Analyses and Responding to Critiques, (December 22, 2008)*
Duncan Bucknell Company (Duncan Bucknell)	1/9/2009
Eli Lilly and Company (Douglas K. Norman)	12/19/2008
Essential Action (Sarah Rimmington)	12/20/2008
Generic Pharmaceutical Association (Lisa K. Layman)	9/30/2008
Hospira, Inc.	5/19/2009
Hospira, Inc. (Lori N. Bowman)	9/30/2008
Hospira, Inc. (Lori N. Bowman)	12/22/2008
Johnson & Johnson, Inc.	3/17/2009 • *Executive Summary,Achieving the Right Balance between Innovation and Competition: The Role of Data Exclusivity*
Johnson & Johnson, Inc.	3/17/2009 • *Achieving the Right Balance between Innovation and Competition: The Role of Data Exclusivity*
Matrix Global Advisors, LLC (Alex M. Brill)	12/22/2008
Momenta Pharmaceuticals, Inc. (Bruce A. Leicher)	9/30/2008
Momenta Pharmaceuticals, Inc. (Bruce A. Leicher)	12/22/2008
Mylan, Inc. (David Rice)	1/5/2009
Novartis Corporation (Robert E. Pelzer)	9/29/2008
Novartis Corporation (Robert E. Pelzer)	12/22/2008
Office of Health Economics (Adrian Towse)	12/31/2008

(Continued)

NAME	TITLE OF COMMENT
Pharmaceutical Care Management Association (Missy Jenkins)	9/26/2008
Pharmaceutical Research and Manufacturers of America (Billy Tauzin)	9/30/2008
Pharmaceutical Research and Manufacturers of America (Billy Tauzin)	12/22/2008
Robert J. Reinhart	10/19/2008
Talecris Biotherapeutics (Bruce Bunyan)	9/30/2008
Teva Pharmaceuticals USA (David T. Sanders)	10/8/2008
Willkie Farr & Gallagher LLP on behalf of Hospira, Inc. (Kelsey I. Nix)	12/22/2008
Winston & Strawn LLP (James F. Hurst,.)	12/22/2008
Wyeth (Matthew D. Eyles)	9/30/2008
Wyeth (Matthew D. Eyles)	12/18/2008
Zuckerman Spaeder LLP (William B. Schultz)	12/22/2008

Primary Source: http://ftc.gov/os/comments/healthcarecompissues/index.shtm

Appendix D: Participants at FTC Roundtable, November 21, 2008

NAME	AFFILIATION
Alexis **Ahlstrom**	Avalere Health LLC
Geoffrey **Allan**	Insmed Incorporated
AaronF. **Barkoff**	McDonnell Boehnen Hulbert & Berghoff LLP
Rachel E. **Behrman**	U.S. Food and Drug Administration
Alex M. **Brill**	American Enterprise Institute; Buchanan, Ingersoll & Rooney PC
Steven B. **Brugger**	Momenta Pharmaceuticals, Inc.
Edward **Buckley**	Biotechnology Industry Organization
Kenneth J. **Dow**	Johnson & Johnson, Inc.
Suzanne B. **Drennon**	Federal Trade Commission
Christopher J. **Garmon**	Federal Trade Commission

(Continued)

NAME	AFFILIATION
David **Golding**	CVS Caremark Corporation
Ken **Goldman**	Novartis Corporation
Mark A. **Goshko**	Teva Pharmaceutical Industries Ltd.
Henry C. **Grabowski**	Duke University
Pamela Jones **Harbour**	Federal Trade Commission
Paul **Heldman**	Potomac Research Group
Linda R. **Horton**	Hogan & Hartson LLP
Elizabeth A. **Jex**	Federal Trade Commission
Esther M. **Kleppinger**	Wilson Sonsini Goodrich & Rosati PC
Jeffrey P. **Kushan**	Sidley Austin, LLP
John A. **Lane**	Hospira, Inc.
Bruce A. **Leicher**	Momenta Pharmaceuticals, Inc.
David A. **Manspeizer**	Wyeth
Suzanne T. **Michel**	Federal Trade Commission
Steven B. **Miller**	Express Scripts, Inc.
Douglas K. **Norman**	Eli Lilly and Company
Naomi **Pearce**	Hospira, Inc.
Audrey **Phillips**	Johnson & Johnson, Inc.
Hans **Sauer**	Biotechnology Industry Organization
William B. **Schultz**	Zuckerman Spaeder LLP
Rochelle K. **Seide**	Schwegman, Lundberg & Woessner PA
Christine J. **Siwik**	Rakoczy Molino Mazzochi Siwik LLP
Mateja **Urlep**	Novartis AG (Sandoz Ltd.)
Michael S. **Wroblewski**	Federal Trade Commission
Bryan C. **Zielinski**	Pfizer, Inc.

Primary Source: http://ftc.gov/bc/workshops/hcbio/agenda/fobagenda.pdf

End Notes

[1] These sales figures are based on wholesale prices reported in the IMS Top Line Industry Data. Press Release, IMS Health, IMS Health Reports U.S. Prescription Sales Jump 3.8 Percent in 2007, to $286.5 Billion (March 12, 2008), *available at* http://www.imshealth.com (follow "Press Room" hyperlink; then follow "IMS Health Care Reports News Release" hyperlink).

[2] *See* The Federal Food, Drug, and Cosmetic Act, 21 U.S.C.A. § 301 *et seq.* (2009), as amended by the Drug Price Competition and Patent Term Restoration Act of 1984 (the "Hatch-Waxman Act" or "Hatch-Waxman") and the Medicare Prescription Drug, Improvement, and Modernization Act of 2003, 21 U.S.C.A. § 355(j) (2009) and 35 U.S.C.A. § 271(e) (2009) [United States Code Annotated].

[3] *See generally* Jennifer S. Haas, et al., *Potential Savings from Substituting Generic Drugs for Brand-Name Drugs: Medical Expenditure Panel Survey, 1997-2000*, 142 ANNALS INTERNAL MED. 891 (June 2005); Wendy H. Schacht and John R. Thomas, Congressional Research Service (CRS), Library of Congress, *Report for Congress, Follow-On Biologics: Intellectual Property and Innovation Issues,* at 4 & 18, 110th Cong. (Jan. 17, 2008), *available at* http://www.biosimilars.com/CRS_FOBs.pdf.

[4] *See How Pay-for-Delay Settlements Make Consumers and the Federal Government Pay More for Much Needed Drugs: Hearing Before the H. Subcomm. on Commerce, Trade, and Consumer Protection, Comm. on Energy and Commerce*, 111th Cong. (2009) (Prepared Statement of the Federal Trade Commission), *available at* http://www.ftc.gov/os/2009/03/P859910payfordelay.pdf.

[5] Immunogenicity raises safety and effectiveness concerns because of a biologic drug's ability to stimulate an immune response. *See* Letter from Frank M. Torti, Principal Deputy Comm'r and Chief Scientist, FDA, to Frank Pallone, Jr., Chmn., H. Subcomm. on Health, (Sept. 18, 2008) at 1, *available at* http://energycommerce.house.gov/Press_110/fdabiosimilarrespons20080918.pdf.

[6] *See* H.R. 1427, 111th Cong. § 3(a) (2009); H.R. 1548, 111th Cong. § 101 (2009).

[7] The Commission has reviewed pharmaceutical and biotechnology mergers for over 30 years, and has conducted numerous investigations and enforcement actions involving the conduct of branded and generic small-molecule drug manufacturers arising in the context of the Hatch-Waxman Act. *See* http://www.ftc.gov/be/0608rxupdate.pdf. The Commission also conducted a detailed empirical study of the experience during the 1993-2001 under the Hatch-Waxman Act's procedures designed to facilitate entry of generic drugs. Federal Trade Commission, *Generic Drug Entry Prior to Patent Expiration: An FTC Study* (July 2002). Since 2004, FTC staff has reviewed every drug company patent settlement filed under Hatch-Waxman, and issued annual reports on the types of patent settlements being undertaken. The reports are available at http://www.ftc.gov/be/healthcare/drug/index.htm.

[8] Historically, some biologic protein products have been regulated as drugs under the FD&C Act, including insulin, and human growth hormones. The FDA has approved six follow-on protein products under the FD&C Act. *See Assessing the Impact of a Safe and Equitable Biosimilar Policy in the United States Hearing Before H. Subcomm. on Health and the H. Comm. on Energy and Commerce,* 110th Cong. (2007) (statement of Janet Woodcock, Deputy Comm'r, Chief Medical Officer, FDA), *available at* http://energycommerce.house.gov/cmte_mtgs/110-he-hrg.050207.Woodcock-testimony.pdf.

[9] F.M. Scherer & David Ross, INDUSTRIAL MARKET STRUCTURE AND ECONOMIC PERFORMANCE, 3d Ed. 621 (1990).

[10] *See* Joseph DiMasi & Cherie Paquette, *The Economics of Follow-on Drug Research and Development*, 22 PHARMACOECONOMICS Supp 2:1-14, 10 (2004). Although this study examined pharmaceutical products primarily, it included several biologic drugs as well.

[11] *See* Organization for Economic Co-operation and Development, *Challenges Obstacles Faced by Competition Authorities in Achieving Greater Economic Development Through the Promotion of Competition, Contribution from the United States* (Feb. 5, 2004), *available at* http://www.scribd.com/doc/1170395/US-Federal-Trade-Commission-2004Challenges20Obstacles20aced20by20Competition.

[12] This chapter uses the term "branded exclusivity" rather than "data exclusivity" because current legislative proposals permit an FOB applicant to rely on FDA's finding or conclusion that an approved pioneer drug is safe and effective. This reliance does not involve disclosure to the FOB applicant, or to the public, of the data in the pioneer manufacturers' application. *See* Letter from Director Steven K. Glason, Center for Evaluation and Research ("CDER"), FDA to Petitioners (May 30, 2006) at 6, *available at* http://www.fda.gov/ohrms/dockets/dockets/04P0231/04P-0231-pdn0001.pdf. The term "data exclusivity" suggests a use of the information that is inconsistent with FDA's longstanding interpretation of its approval process.

[13] Henry C. Grabowski, *Follow-on Biologics: Data Exclusivity and the Balance Between Innovation and Competition*, 7 NATURE REVIEWS DRUG DISCOVERY 479, 483 (June 2008).

[14] *See* David Reiffen & Michael Ward, *"Branded Generics" As A Strategy To Limit Cannibalization of Pharmaceutical Markets, 28* MANAGERIAL AND DECISION ECONOMICS, 251-265, 264 (2005), *available at* http://ftc.gov/be/healthcare/wp/12_Reiffen_BrandedGenericsAsAStrategy.pdf.

[15] *See* FDA, Center for Drug Evaluation and Research, 180-Day Generic Drug Exclusivity (2001), *available at* http://www.fda.gov/cder/about/smallbiz/generic_exclusivity.htm#COURT ("This 180-day exclusivity provision has been the subject of considerable litigation and administrative review in recent years…").

[16] *See How Pay-for-Delay Settlements Make Consumers and the Federal Government Pay More for Much Needed Drugs: Hearing Before the H. Subcomm. on Commerce, Trade, and Consumer Protection, Comm. on Energy and Commerce*, 111th Cong. (2009) (Prepared Statement of the Federal Trade Commission), *available at* http://www.ftc.gov/os/2009/03/P859910payfordelay.pdf.

[17] The Commission outlined its preliminary views on the likely effects of an abbreviated regulatory approval pathway for biologic drug products in May 2008. *See* Letter of the Federal Trade Commission to the Honorable Frank Pallone, Jr., Chairman, Subcommittee on Health, Committee on Energy and Commerce, House of Representatives (May 2, 2008), *available at*: http://energycommerce.house.gov/Press_110/110-ltr.050208.respto040308.FTC.pdf.

[18] It is beyond the scope of this chapter to determine whether a 20-year patent life is the optimal period to incentivize innovation in this and other industries that rely on patent protection.

[19] *See* FEDERAL TRADE COMM'N, TO PROMOTE INNOVATION: THE PROPER BALANCE OF COMPETITION AND PATENT LAW AND POLICY (2003), Ch. 3 at 1, *available at* http://www.ftc.gov/os/2003/10/innovationrpt.pdf.

[20] *See* Charles E. Phelps, *Managing the Market: Regulation and Technical Change in Health Care*, HEALTH ECONOMICS, at 498-546 (3rd ed. 2003).

[21] *See* Iain M. Cockburn, *The Changing Structure of the Pharmaceutical Industry*, 23 HEALTH AFFAIRS 1:10-22, 14 (2004).

[22] Biologic drugs are derived from living matter or manufactured in living cells using recombinant DNA biotechnologies. *See* FDA Center for Biologic Drug Evaluation and Research (CBER), *Frequently Asked Questions About Therapeutic Biologic Drug Products, available at* http://www.fda.gov/cder/biologics/qa.htm.

[23] These sales figures are based on wholesale prices reported in the IMS Top Line Industry Data. Press Release, IMS Health, IMS Health Reports U.S. Prescription Sales Jump 3.8 Percent in 2007, to $286.5 Billion (March 12, 2008), *available at* http://www.imshealth.com (follow "Press Room" hyperlink; then follow "IMS Health Care Reports News Release" hyperlink); *see also* CONG. BUDGET OFFICE ("CBO") 110TH Cong., BUDGET OPTIONS VOL.1: HEALTH CARE at 126-28 (2008), *available at* http://cbo.gov/ftpdocs/99xx/doc9925/12-18-HealthOptions.pdf [hereinafter, "BUDGET OPTIONS"].

[24] *See* The Federal Food, Drug, and Cosmetic Act, 21 U.S.C.A. § 301, *et seq.* (2009), as amended by The Drug Price Competition and Patent Term Restoration Act of 1984 [hereinafter, the "Hatch-Waxman Act" or "Hatch-Waxman"] and the Medicare Prescription Drug, Improvement, and Modernization Act of 2003, 21U.S.C.A. § 355(j) (2009) and 35 U.S.C.A. § 271(e) (2009). *See* Appendix B for a description of the new and abbreviated drug approval processes.

[25] *See generally* Jennifer S. Haas *et al.*, *Potential Savings from Substituting Generic Drugs for Brand-Name Drugs: Medical Expenditure Panel Survey, 1997-2000*, 142 ANNALS INTERNAL MED. 891 (June 2005); Wendy H. Schacht & John R. Thomas, Congressional Research Service (CRS), Library of Congress, *Report for Congress, Follow-On Biologics: Intellectual Property and Innovation Issues,* at 4 & 18, 110th Cong. (Jan. 17, 2008), *available at* http://www.biosimilars.com/CRS_FOBs.pdf.

[26] *See Assessing the Impact of a Safe and Equitable Biosimilar Policy in the United States Hearing Before H. Subcomm. on Health and the H. Comm. on Energy and Commerce,* 110th Cong. (2007) (statement of Janet Woodcock, Deputy Comm'r, Chief Medical Officer, FDA), *available at* http://energycommerce.house.gov/cmte_mtgs/110-he-hrg.050207.Woodcock-testimony.pdf [hereinafter, "Woodcock Statement"]. Historically, some biologic protein products have been regulated as drugs under the FD&C Act. The FDA has approved six follow-on protein products under the FD&C Act, including Hylenex (hyaluronidase recombinant human), Hydase (hyaluronidase), Fortical (calcitonin salmon recombinant) Nasal Spray, Amphadase (hyaluronidase), GlucaGen (glucagon recombinant for injection), and Omnitrope (somatropin [rDNA origin]). *Id.*

[27] *See* Bernstein Research Comment (9/29/08) at 2; Biotechnology Industry Organization ("BIO"), Health Overview, *available at* http://www.bio.org/healthcare (last accessed June 8, 2009); CBO, BUDGET OPTIONS at 126; Hospira (Wilkie Farr) Comment (12/22/08) at 5 and Attachment 1. Patent expiration information was obtained from SEC form 10-K filings. FDA maintains a searchable catalog of approved drug products including drug approval history. *See,* Drugs@FDA, *available at* http://www.accessdata.fda.gov/Scripts/cder/DrugsatFDA.

[28] Immunogenicity raises safety concerns because of a biologic drug's ability to stimulate an immune response. An immune response to a therapeutic protein can range from development of detectable but not clinically significant antibodies to an immune response with significant impact on safety or effectiveness, including the potential to decrease or block the clinical effect of the therapeutic protein. *See* Letter from Frank M. Torti, Principal Deputy Comm'r and Chief Scientist, FDA, to Frank Pallone, Jr., Chmn., H. Subcomm. on Health, (Sept. 18, 2008) at 1, *available at* http://energycommerce.house.gov/Press_110/fdabiosimilarrespons20080918.pdf.

[29] *Id.* at 4; Woodcock Statement at 1 ("[T]he idea of *sameness*, as the term is used in the generic drug approval process under the [FD&C] Act and applied to small-molecules, will not usually be appropriate for more structurally complex molecules of the type generally licensed as biological products under the [PHS] Act.").

[30] *See* H.R. 1427, 111th Cong. § 3(a) (2009); H.R. 1548, 111th Cong. § 101 (2009).

[31] *See* Notice of Public Workshops and Roundtables and Opportunity for Comment, Emerging Health Care Competition and Consumer Issues, 73 Fed. Reg. 51479-51482 (Sept. 3, 2008), *available at* http://www.ftc.gov/bc/workshops/hcbio/index.shtml.

[32] *See* 42 U.S.C.A. § 262; 21 U.S.C.A. § 321, *et seq.* (2009).

[33] *See, e.g.*, Ernst R. Berndt *et al.*, *Opportunities for Improving the Drug Development Process: Results from a Survey of Industry and the FDA* (Nat'l Bureau of Econ. Research, Working Paper No. W11425, 2005).

[34] 42 U.S.C.A. § 262; 21 U.S.C.A. § 321, *et seq.*; 21 C.F.R. 601.2; 21 C.F.R. 312 (2009).

[35] *See* Henry Grabowski *et al.*, The Effect on Federal Spending of Legislation Creating a Regulatory Framework for Public Health Service Act Follow-on Biologics: Key Issues and Assumptions, White Paper (July 1, 2007) at 8, 25, 27-28, 33 (unpublished paper on file with Analysis Group, Inc.), *available at* http://www.analysisgroup.com/analysisgroup/News_Study-Effects-Federal-Spending-Follow-On-Biologics-Legislation.aspx [hereinafter "White Paper"].

[36] 21 U.S.C.A. § 355(j)(2)(A)(iv)(2009). Bioequivalence means that the rate and extent of absorption of the generic drug is not significantly different from the rate and extent of absorption of the reference listed drug when administered at the same dosage.

[37] CBO, *How Increased Competition from Generic Drugs Has Affected Prices and Returns in the Pharmaceutical Industr,* at ix (1998), *available at* http://www.cbo.gov/ftpdocs/6xx/doc655/pharm.pdf [hereinafter, "*Increased Competition from Generic Drugs*"].

[38] Testimony of Rachel Berhman, Associate Comm'r for Clinical Programs, Director of the Office of Critical Path Programs, FDA, at FTC Roundtable: Emerging Healthcare Competition and Consumer Issues (Nov. 21, 2008) at 10-20, *available at* http://www.ftc.gov/bc/workshops/hcbio/transcripts/081121biologic-transcript.pdf [hereinafter transcript cites are referenced as [last name] at [page]]; Woodcock Statement at 8-9.

[39] Behrman at 10-20; Rachel Behrman, *Follow-on Biologics: A Brief Overview at FTC Roundtable: Emerging Healthcare Competition and Consumer Issues* (Nov. 21, 2008) at 6 [hereinafter "Behrman Presentation"].

[40] *See* Woodcock Statement at 8 ("Because of the variability and complexity of protein molecules, current limitations of analytical methods, and the difficulties in manufacturing a consistent product, it is unlikely that, for most proteins, a manufacturer of a follow-on protein product could demonstrate that its product is identical to an already approved product."); Behrman Presentation at 6; Behrman at 13-20.

[41] *See* Behrman at 12-13 ("[P]roteins [biologics] . . . are chains of amino acids . . .they can range from very simple to extremely complex, and when they're very complex, they are folded; they have things stuck on them; they can unfold again; and then they can aggregate."); *see also* Norman at 153 ("[T]he chemical [small-molecule] compound itself is something that always looks like chicken wire, so it's got a methyl on one end and maybe an ethyl on the other, but it's going to look like methyl ethyl chicken wire, and every follow-on generic or branded firm] that makes that molecule … is going to make methyl ethyl.").

[42] *See* S. 1695, 110th Cong. (2008); H.R. 1427, 111th Cong. (2009).

[43] *See* Woodcock Statement at 11 ("When the mechanism of action is well understood and there is a significant amount of clinical experience with a product, it may be easier to make a scientific assessment of the ability to rely on conclusions about safety and efficacy from a prior application.").

[44] For a description of the FDA clinical requirements required to approve the first biosimilar product in the U.S. *see* Letter from Director Steven K. Glason, Center for Evaluation and Research ("CDER"), FDA to Petitioners (May 30, 2006) at 7, 25 (Novartis' application for Omnitrope included "CMC[chemistry, manufacturing and control], nonclinical pharmacology and toxicology, human pharmacokinetic and pharmacodynamic, and clinical safety and effectiveness data," including 3 Phase III trials), *available at* http://www.fda.gov /ohrms/dockets/dockets/04P0231/04P-0231-pdn0001.pdf [hereinafter "FDA's Second Response to Omnitrope CPs"]; *see* Grabowski, *White Paper* at 25-26 ("Obtaining approvable [FOB] manufacturing capacity may take 3 to 7 years.").

[45] *See* Henry Grabowski *et al., Entry and Competition in Generic Biologics*, 28 MANAGERIAL AND DECISION ECONOMICS 439-51 (development time for FOB estimated at 5-8 years, 3 years for preclinical work, 2-4 years for clinical trials and 1 year for FDA approval); Grabowski, *White Paper* at 25-26 ("FOB development and trials will likely take 3 to 5 years, and obtaining FDA approval another one and a half to two years."); *see also id.* at 5, 27-30.

[46] The term "interchangeable" is not currently defined in the PHS Act. Many panelists and commenters suggested that interchangeability was unlikely to be possible in the near term. *See* Buc kley at 47 ("In Europe, to date 14 countries have ruled that these products are not interchangeable"); *see id.* at 51; Phillips at 103. Participants noted that the European Union ("EU") member states (including France, Germany, United Kingdom, Italy and Spain) have all rejected the practice of substitution of a biologic by the pharmacist without the physician's consent. Amgen Comment (9/30/08) at 2-3, 6; Novartis Comment (9/29/08) at 2, 16-17; Brugger at 38-39.

[47] *See* Behrman at 78; *see also* Momenta Comment (12/22/2008) at 3 (new analytic tools developed by Momenta to characterize proteins may provide significant "value and cost savings to the innovator drug development process . . . to enhance the quality of their products by more precisely controlling variability of a number of attributes in the final drug product . . . and reduce the need for very costly, potentially unnecessary clinical trials.").

[48] Novartis Comment (9/29/08) at 3-4; Brugger at 54 ("We've developed an innovative analytical approach to these complex molecules, both in better understanding the [biologic] product, but also a deeper understanding of the manufacturing process.*"); see also id.* at 55, 79; Momenta Comment (9/30/08) at 2.

[49] *See* FEDERAL TRADE COMM'N, PHARMACY BENEFITS MANAGERS: OWNERSHIP OF MAIL ORDER PHARMACIES, (August 2005) [hereinafter "FTC PBM REPORT"] at 63, *available at* http://www.ftc.gov/reports/ pharmbenefit05/050906pharmbenefitrpt.pdf.

[50] Generally, each PBM negotiates with branded drug manufacturers for discounts or market share payments that are based on the branded drug's preferred status on the PBM's drug formulary or on the branded drug's market share among the PBM's members. Branded drug manufacturers make these payments to encourage the PBM to dispense their branded drugs rather than competing branded products within a therapeutic class. Drug formularies are used primarily for drugs dispensed in a retail pharmacy environment. *See* FTC PBM REPORT, Ch. 1 at 4, 6.

[51] Joseph A. DiMasi & Cherie Paquette, *The Economics of Follow-on Drug Research and Development*, 22 PHARMACOECONOMICS Supp 2:1-14 (2004) (The study included several biologic drugs). In the 1990s all of the breakthrough products had branded competitors in clinical development at or before their approval; *id.* at 10.

[52] *Id.*; F.M. Scherer, *Markets and Uncertainty in Pharmaceutical Development* 13 (FACULTY RESEARCH WORKING PAPERS SER., HARV.UNIV., JOHN F. KENNEDY SCHOOL OF GOV'T, 2007), *available at* http://ksgnotes1.harvard.edu/Research/wpaper.nsf/rwp/RWP07-039/$File/rwp_07_039_scherer.pdf.

[53] *See* FEDERAL TRADE COMM'N, TO PROMOTE INNOVATION: THE PROPER BALANCE OF COMPETITION AND PATENT LAW AND POLICY (2003), CH. 2 AT 11 [HEREINAFTER "FTC PATENT REPORT"], *available at* http://www.ftc.gov/os/2003/10/innovationrpt.pdf; FTC/DOJ HEALTH CARE REPORT, CH. 6 AT 3-9.

[54] Although the competing branded product's list price, including Average Wholesale Price ("AWP") or Wholesale Acquisition Cost ("WAC"), is typically at parity, the firms compete by offering price discounts to the largest, most sophisticated, and price sensitive customers, such as PBMs. These discounts are confidential. *See* IMPROVING HEALTH CARE: A DOSE OF COMPETITION: A REPORT BY THE FEDERAL TRADE COMM'N AND THE DEP'T OF JUSTICE (July 2004), Ch. 7, at 11-17, *available at* http://www.ftc.gov/reports/healthcare/040723healthcarerpt.pdf [hereinafter "FTC/DOJ HEALTH CARE REPORT"]; *see also* DiMasi and Paquette, *The Economics of Follow-on Drug Research and Development,* at 12 (average discount offered by subsequent branded rivals were 26% off price leader and 14% off the class average); CBO, *Increased Competition from Generic Drugs,* at 24-25.

[55] *See* FTC/DOJ HEALTH CARE REPORT, CH. 6-7; FTC PATENT REPORT, Ch. 2, at 11.

[56] In general, if the patent application was filed after June 7, 1995, the patent expires 20 years from the date on which the application was filed. 35 U.S.C.A. § 154(a)(2)-(3) (2009). If the application was filed by June 7, 1995 and issued after June 7, 1978, the term is the later of 17 years from issuance or 20 years from filing. 35 U.S.C.A. § 154(c). If the application was filed by June 7, 1995 and issued before June 8, 1978, the expiration date was 17 years from issuance, i.e., 1995 or earlier.

[57] Generally, the number of generic entrants increases with the market size. In one study of the 40 oral small-molecule drugs with patent expiry between 1992 to1998, an average of 12 generic firms entered when the market size before patent expiry was over $250 million. In comparison, when market size was less than $250 million, only 5 generic firms entered. Grabowski, *Entry and Competition*, at 440, 444-46; *see also* David Reiffen & Michael R. Ward, *Generic Drug Industry Dynamics,* 87 REVIEW OF ECONOMICS AND STATISTICS, 37–49, 38 (2005) ("more firms enter, and enter more quickly, in markets with greater expected rents"), *available at* http://www.ftc.gov/be/econwork.htm.

[58] *See* Reiffen, *Generic Drug Industry Dynamics*; *see also* CBO, *Increased Competition from Generic Drugs,* at 28; Grabowski, *Entry and Competition*, at 444, 446 (economic analysis concludes that 1 generic entrant results in discounts of 10%, 5 generic firms 37%, 5 entrants 40%, 10 entrants 60%, and 95% after 20 entrants); Grabowski, *White Paper* at 42-44, 52-53; OTA, *Pharmaceutical R&D,* at 297; Roy Levy, THE PHARMACEUTICAL INDUSTRY: A DISCUSSION OF COMPETITIVE AND ANTITRUST ISSUES IN AN ENVIRONMENT OF CHANGE (1999) at 73-76, 197 [hereinafter "LEVY REPORT"], *available at* http://www.ftc.gov/os/2006/07/P052103BarrierstoGenericEntryTestimonySenate07202006.pdf.

[59] *See* Grabowski at 42 (generic erosion 90%), Heldman at 26-28 (generic erosion 80%); *see also* Golding at 49, Buckley at 52; Grabowski, *Entry and Competition*, at 444 (brands lose 67% of market share within a year); CBO, *Increased Competition from Generic Drugs* at 29-31; David Reiffen & Michael Ward, *"Branded Generics" As A Strategy To Limit Cannibalization of Pharmaceutical Markets, 28* MANAGERIAL AND DECISION ECONOMICS, 251-65 (2005), *available at* http://ftc.gov/be/healthcare/wp/12_Reiffen_BrandedGenericsAsAStrategy.pdf; Henry Grabowski & John Vernon, *Longer Patents for Increased Generic Competition in the U.S.: The Hatch-Waxman Act After One Decade*, 10 PHARMACOECONOMICS supp. 2:110-23 (1996) (brands lost 50% of prescriptions within a year); U.S. Cong., Office of Technology Assessment ("OTA"), *Pharmaceutical R&D: Costs, Risks and Rewards*, OTA-H-522 (1993), at Table F-3, p. 297, *available at* http://www.princeton.edu/~ota (Princeton University. hosts the OTA legacy site, follow "OTA publications" hyperlink and use search engine there to find article by title) [hereinafter "*Pharmaceutical R&D"*].

[60] *See* CBO, *Research and Development in the Pharmaceutical Industry* at 48 (2006), *available at* http://www.cbo.gov/ftpdocs/76xx/doc7615/10-02-DrugR-D.pdf; CBO, *Increased Competition from Generic Drugs,* at 27-30; Grabowski, *Entry and Competition*, at 444-45.

[61] *See* FTC PBM REPORT at x, 12, 74-75.

[62] *See* Reiffen & Ward, *Branded Generics*; Grabowski & Vernon, *Longer Patents for Increased Generic Competition*; Grabowski, *Entry and Competition*, at 444.

[63] *See* Duke University Comment (12/23/08) at Table 3; BIO Comment (9/30/08) at 2; Grabowski *White Paper* at 2, 4-6, 8-9, 40-41, 48; *see also* Paul Heldman *et al.*, Citigroup Research, Citigroup Global Markets, *A Global "Generic Biologics" Guidebook* at 5 (November 6, 2006)[hereinafter "Citigroup 2006 FOB Guidebook"]; *Safe and Affordable Generic Biotech Drugs: The Need for a Generic Pathway: Hearing before the H. Oversight and Gov't Reform Comm.*, 110th Cong. 1-14 (2007) (statement of Henry Grabowski, Duke University), *available at* http://oversight.house.gov/documents/20070416132526.pdf; Grabowski, *Entry and Competition*, at 448-49 (after extensive economic modeling, the authors conclude that and FOB prices relatively close in price to branded biologics); CBO Cost Estimate (S.1695), *Biologics Price Competition and Innovation Act of 2007 S. 1695, As Ordered Reported by the S. Comm. on Health, Education, Labor, and Pensions on June 27,*

2007 (June 25, 2008), *available at* http://www.cbo.gov/ftpdocs/94xx/doc9496/s1695.pdf [hereinafter "S. 1695 Report"].

[64] *See* CBO S. 1695 Report at 6 ("CBO expects that certain drugs could face competition from several firms by 2018, although we believe it would be more typical for an innovator biologic to face competition from between one and three competitors."); Grabowski, *Entry and Competition*, at 446-47 (because of bioreactor capacity constraints and high fixed costs for *de novo* biologic manufacturing facilities, the number of FOB entrants is likely to be smaller than that predicted for generic small-molecule markets for the foreseeable future); *see also* Buckley at 53 ("The number of entrants will certainly be fewer. .. There are technological know-how [barriers] . . . the price of clinical trials ... the length of the approval process, the likelihood of a successful application . . .and you start to see that the number of players that can submit a successful application is just much smaller."); Amgen Comment (9/30/08) at 5; Alhstrom at 44-45; Grabowski at 42; Heldman at 25; Lane at 46. The technological barriers to entry vary on the complexity of the biological product. Several FOB manufacturers are predicted to be able to obtain FDA approval for biosimilar versions of first generation recombinant proteins. However, as the biological products become more scientifically complex, as in the case of many of the monoclonal antibodies, the technological barriers to entry are so significant that few predict FOB in the next decade.

[65] *See* Sumanth Kambhammettu, Senior Research Analyst, Frost & Sullivan, *The European Biosimilars Market: Trends and Key Success Factors*, (Oct. 27, 2008)("average cost of bringing a biosimilar to market is around $100-$200 million"), http://www.obbec.com/specialreports/20-biopharmaceuticals/2152-the-european-biosimilars-market-trends-and-key-success-factors; CBO S. 1695 Report at 6; Duke University Comment (12/23/08) at Table 3; BIO Comment (9/30/08) at 2; Grabowski *White Paper* at 2, 4-6, 8-9, 40-41, 48; Grabowski, *Entry and Competition*, at 442; Citigroup 2006 FOB Guidebook at 5; *see also* Ahlstrom at 53; Lane at 40, 46; Zuckerman Comment (12/22/08) at 12.

[66] *See* BIO Comment (9/30/08) at fn. 2, 1, 9, 17, 20; Grabowski, *White Paper*; CBO S. 1695 Report at 4-7; GPhA Comment (9/30/08) at 3 (citing CBO, *Increased Competition from Generic Drugs).*

[67] *See* Novartis Comment (9/29/08) at 7; Wyeth Comment (12/18/08) at 6 ("[T]he cost of manufacturing facilities is staggering, and this large investment must be made long before a product is approved by the regulatory agencies."); Henry Grabowski, *Follow-on Biologics: Data Exclusivity and the Balance Between Innovation and Competition*, 7 NATURE REVIEWS DRUG DISCOVERY 479, 483 (June 2008) [hereinafter "NATURE"]; *Patent Reform Act of 2007: Hearing on H.R.. 1908 Before the H. Subcomm. on Courts, the Internet, and Intellectual Property of the H. Comm. on Judiciary,* 110th Cong. 65 (2007) (statement of Kevin Sharer, Amgen, Inc. ("Amgen" CEO), *available at* http://judiciary.house.gov/hearings/April2007/ Sharer070426.pdf. ("It takes approximately 5 years and $1 billion to build a factory to produce biotech medicines."). FOB manufacturers will likely have to develop their own in-house manufacturing because the worldwide capacity constraints puts greater leverage in the hands of contract manufacturers to seek maximum profits by maintaining their relationships with branded firms and the highest revenue producing branded products. Grabowski, *Entry and Competition*, at 442.

[68] *See* Lane at 35-36; Urlep at 34.

[69] Janet Woodcock *et al.*, *The FDA's Assessment of Follow-On Protein Products: a Historical Perspective*, 6 NATURE REVIEWS DRUG DISCOVERY 437-42 (June 2007); Grabowski, *Entry and Competition*, at 446 (only 3 entrants predicted in markets where the branded biologic has sales over $1 Billion).

[70] *See* Natasha Singer, *Bristol-Myers's Reliance on Three Drugs Casts Doubt on Strategy*, N.Y. TIMES, Jan. 27, 2009, at B3 (listing recent biopharmaceutical acquisitions); Andrew Jack, *Sanofi-Aventis Ready to Join Pfizer on the Acquisition Trail,* FINANCIAL TIMES, February 2, 2009 at 13 ("[T]he latest is a spurt of consolidation in the pharmaceuticals sector"); Andrew Pollack, *Wyeth Deal May Slow Pfizer Biotech Acquisitions* , N.Y. TIMES, JAN. 26, 2009, at B4 ; ("Schering-Plough's considerable biologics expertise will complement Merck's novel proprietary biologics platform"), *available at* http://www.merck.com/newsroom/press_releases/corporate/2009_0309.html.

[71] *See* CVS Caremark Comment (12/22/08) at 3; Momenta Comment (12/22/2008) at 4; Andrew Jack, *AstraZeneca Chief Calls the Shots*, FINANCIAL TIMES, December 23, 2008 at 18 ("The move is the third instance in recent weeks of a large pharmaceutical company [AstraZeneca] that has been traditionally focused on developing innovative medicines to express a desire to shift to generic [biologic] medicines."); Susan Todd, *Merck Launches Biologic Division Drugmaker to Invest $1.5 B into Venture*, NEWARK STAR LEDGER, December 10, 2008 at 61 ("We anticipate that [Merck] will take a leadership position in follow-on biologics."). Merck BioVentures is developing an FOB equivalent to Amgen's Aranesp which Merck expects to launch in 2012 to be followed by five more FOBs of several best-selling biologic drugs due to lose patent protection by 2017. On November 24, 2008, Eli Lilly and Company acquired the biologic company Imclone, the biologic manufacturer of Erbitux (with global sales of $3.6 billion) for $6.5 billion beating out Bristol-Myers Squibb. Press Release, Eli Lilly & Co., Lilly to Acquire Imclone Systems in $6.5 Billion Transaction, (October 6, 2008), *available at* http://newsroom.lilly.com/releasedetail.cfm?ReleaseID+338523. In June 2007, AstraZeneca completed its $15 billion acquisition of MedImmune. AstraZeneca, Annual Report and Form 20-F Information, at 83-84 (2007); Bernstein Research Comment (9/29/08) at 2.

[72] *See, e.g.*, CCPM Comment (9/30/08) at 3 ("In the absence of a designation as interchangeable, it likely will take longer for the [biosimilar] to garner significant market share and brand manufacturers will have less incentive to compete based on price."); *see also* CVS Caremark Comment (12/22/08) at 4.

[73] *See, e.g.*, Hospira (Wilkie Farr) Comment (12/22/08) at 5 ("Without an "interchangeable designation, biosimilar companies would be compelled to invest significant sums to market and promote biosimilars, thus driving up the cost to the consumer. Reference companies also would have less incentive to compete on price. Reference drug companies would be more likely to try to out-market the biosimilar companies, further driving up the costs of *both* the reference drug and market entry by the biosimilar.").

[74] *See supra* notes 35-40 and accompanying text regarding negotiated price discounts for different purchasers.

[75] *See* Amgen Comment (9/30/08) at 2-3; BIO Comment (12/22/08); Ahlstrom at 43.

[76] Generally, the FDA approves the use of the same name for a generic small-molecule as the reference branded drug because both products share the same the active ingredient. In contrast, an FOB drug manufactured by a different process than the reference branded biologic drug may share the same mechanism of action, may share the same efficacy and side effects, and may even be considered or approved as interchangeable with the reference branded biologic drug but may still not be given the same name as the brand. *See* Horton at 98; BIO Comment (9/30/08) at 4; PCMA Comment (9/26/2008) at 5; FDA Considerations: Discussion by National Regulatory Authorities with World Health Organization (WHO) On Possible International Non-proprietary Name (INN) Policies for Biosimilars (Sept 1, 2006) ("The world community may ultimately decide that INN policy for this class of products should be treated differently than that for small-molecule drugs."), *available at* http://www.fda.gov/cder/news/biosimilars.htm.

[77] CVS Caremark Comment (12/22/08) at 4 ("Given the uncertainty surrounding the equivalence of innovator and follow-on biologics, PBM, payors and physicians are more likely to be focused on clinical information and dialogue about the prudence of switching to a particular follow-on biologic or innovator drug. This ad hoc, non-uniform approach will ultimately drive the adoption of follow-on biologics, but at a pace than seen with generic small-molecule drugs.").

[78] *Id.*

[79] CBO S. 1695 Report; Ahlstrom at 43.

[80] Golding at 64-65; CVS Caremark Comment (12/22/08) at 1. Specialty pharmaceuticals often are distributed in separate channels to preserve the viability and safe administration of the products.

[81] Golding at 64-65.

[82] For example, costs for senior citizens for biologic drugs are generally reimbursed under Medicare Part B, rather than Part D. *See* CBO BUDGET OPTIONS at 106, 126-27; AARP Comment (12/22/08) at 1; CVS Caremark Comment (12/22/08) at 7.

[83] Alhstrom at 43-45 (noting that insurance plans and PBMs immediately cover generic drugs and immediately implement tools to switch their patients from the brand to the generic small-molecule drug which results in the 80-90% share shift to generics a market dynamic that she does not predict will be duplicated in the biologic-FOB market experience any time in the near future."); Buckley at 47 ("It's going to be the decision of the physician and the patient as to whether or not a drug will be substituted for a therapy that they may already be on or a therapy that they may be considering taking."); *see also* Golding at 49.

[84] Mylan Comment (1/5/09) at 5-6. In contrast to the authority CMS has to incentivize the use of generic small-molecule products, currently, there is no express statutory authority for the CMS to reimburse FOBs in such a way as to incentivize utilization of the lowest priced biologic product. CVS Caremark Comment (12/22/08) at 7; Miller at 213-14. The Congressional Budget Office has indicated that a change to the Medicare Part B reimbursement methodologies would be needed to maximize savings from FOB products. *See generally* CBO BUDGET OPTIONS; CVS Caremark Comment (12/22/08) at 7; Heldman at 29-31 ("The current formula under Medicare provides a financial incentive for physicians and hospitals, when using the drugs in an outpatient setting to use the higher cost drugs...because Medicare reimburses at the average sales plus a 6 percent markup. In addition, current law requires Medicare [to give] a follow-on biologic that the FDA doesn't deem interchangeable . . . a separate billing code…."); Amgen, Inc. v. F. Hoffmann-La Roche Ltd.., 2008 U.S. Dist. LEXIS 77343 at *169-73 (D. Mass. Oct. 2, 2008) (noting that the court could not conclude that entry by Roche's branded EPO biologic drug, Mircera, would reduce Medicare Part B reimbursement for EPO drugs).

[85] CVS Caremark Comment (12/22/08) at 4; Buckley at 52-53; Bernstein Research Comment (9/29/08) at 1-2; BIO Comment (9/30/08) at fn. 2.; Amgen Comment (9/30/08) at 5 ("The combination of these factors will make it very unlikely that biosimilar products will bring about the price differential that generic products do."); Momenta Comment (9/30/08) at 3 ("The likely competitive effect of a follow-on biologic entering the market is the gradual reduction in prices of the biologic."); Hospira Comment (9/30/08) at 1 ("The best estimate is that the biosimilar EPOs [in the EU] appear to be priced approx 25 - 30 percent below the innovator's price prior to the entry of any biosimilar."); CCPM Comment (9/30/08) at 2 ("According to the March, 2008, edition of the Red Book, Omnitrope's price is a 34% discount from the original product."); Grabowski, *White Paper* at 6 ("The extent of entry will likely be much lower for FOBs than for conventional generic drugs, reflecting differences in market size and high fixed costs of entry for many biologics. Average price effects and rates of FOB uptake for innovator products are likely to be limited in the short run due to the low number and timing

of entry of FOBs, limitations to perceived substitutability between innovator biologics and FOBs on the part of physicians and patients, incentives for limited price-based competition between FOBs and innovator products….”); Grabowski, *Entry and Competition*, at 449.

[86] *See* Grabowski, *White Paper* at 2.

[87] *Id.*; Heldman at 25, 27.

[88] Heldman at 25, 27; Brugger at 39; Urlep at 56 (“We have to invest into primary marketing to overcome this with our data, which we created during the development programs.”); Hospira (Wilkie Farr) Comment (12/22/08) at 2 (“Without an ‘interchangeable’ designation, biosimilar companies would be compelled to invest significant sums to market and promote biosimilars, thus driving up the cost to the consumer.”).

[89] Grabowski at 42.

[90] Brugger at 74.

[91] Linda Horton, *The European Experience with Follow-on Biologics Legislation at FTC Roundtable: Emerging Healthcare Competition and Consumer Issues* (Nov. 21, 2008) at 3 [hereinafter “Horton Presentation”].

[92] *See* Appendix B for a discussion of the statutory authority that permits the FDA to approve a limited number of biosimilar products under the FD&C Act.

[93] Although the E.U. approved biosimilar filgrastim on February 13, 2009, market experience was too limited to include in this chapter. *See* Press Release, Sandoz, Sandoz Receives European Commission Approval for Biosimilar Filgrastim (Feb. 13, 2009), *available at* http://www.sandoz.com/site/en/media_room/press_releases_news/090213.shtml; Hospira Comment (May 11 2009) at 3, Amgen, Q1 2009 Earnings Call, at 19-20 (April 23, 2009). In Europe, the pioneer G-CSF products consist of Amgen’s Neupogen (filgrastim), Amgen’s Neulasta (pegylated filgrastim) and Chugai’s Granocyte (lenograstim), while the biosimilar products consist of Teva’s Tevagrastim, Ratiopharm Ratiograstim and Ratiopharm filgrastim, CT Arzneimittel’s Biograstim, Novartis’ Zarzio (marketed by Novartis’ Sandoz division), and Filgrastim Hexal (marketed by Novartis’ Hexal division).

[94] Hospira, Inc. ("Hospira") and STADA Arzneimittel AG co-market their product Retacrit/Silapo in Germany. Market shares for Novartis' Binocrit, Novartis' Epoetin alpha Hexal and Medice' Arzneimittel Putter GmbH&Co KG' s Abseamed are represented together. On February 17, 2009, Shire plc. ("Shire") discontinued selling Dynepo in Europe for commercial reasons, and its marketing authorization was rescinded in March, 2009. *See* http://www.emea.europa.eu/humandocs/PDFs/EPAR/dynepo/12666909en.pdf.

[95] Bernstein Research Comment (9/29/08); Paul Heldman, *Follow-On Biologic Market: Initial Lessons and Challenges Ahead at FTC Roundtable: Emerging Healthcare Competition and Consumer Issues* (Nov. 21, 2008) at 7 [hereinafter “Heldman Presentation”]; Heldman at 26-27; Lane at 36 (“on a unit basis [we] have actually captured 23 percent of the first gen market”); Novartis Comment (9/29/08) at 2-3. Some of the share estimates differ because some estimates are calculated based on units while others are based on different measures of sales. *Id. See also* Amgen, Q1 2009 Earnings Call, at 19-20 (April 23, 2009).

[96] *See* Amgen, Q1 2009 Earnings Call, at 19-20 (April 23, 2009) (stating that Hospira’s Retacrit and the other biosimilars together account for only 5 percent market share).

[97] For a discussion of the novel issues involved with the approval of Omnitrope, *see* Sandoz, Inc. v. Leavitt, 427 F. Supp. 2d 29 (D.D.C. 2006); *see generally* FDA’s Second Response to Omnitrope CPs; Letter from Janet Woodcock, Director, CDER, FDA to Petitioners (October 14, 2003), *available at* http://www.fda.gov/ohrms/DOCKETS/dailys/03/oct03/102403/03p-0408-pdn0001.pdf [hereinafter “FDA’s First Response to Omnitrope CPs”]. When the FDA approved Sandoz’s Omnitrope on May 31, 2006, it did not rate Omnitrope as therapeutically equivalent to and automatically substitutable for Genotropin. *See* Letter from Paulo Costa, President & CEO, Novartis Corp. to Frank Pallone, Jr., Chmn, and Nathan Deal, H. Subcomm. on Health (May 1, 2008) at 9-10, *available at* http://energycommerce.house.gov/Press_110/110-ltr.050108.respto040308.Novartis.pdf.

[98] Heldman at 23; Citigroup 2006 FOB Guidebook at 2; Bernstein Research Comment (9/29/08) at 12-13.

[99] *Id*; FDA’s Second Response to Omnitrope CPs at 7.

[100] Heldman at 28, Heldman Presentation at 3-6.

[101] *See* Heldman at 28, Heldman Presentation at 3-6; CBO S.1695 Report. Heldman notes that aggressive discounts offered in the market to PBMs and other payors are generally non-public and not captured in the WAC data available from IMS and other sources. Branded firms compete on prices not by lowering the list, WAC or AWP to all customers, but by offering discounts off those prices to the most price sensitive, sophisticated,and largest purchasers. *See* Heldman at 24-25; FTC PBM REPORT at 48-54; LEVY REPORT at 183; *see* generally CBO, *Research and Development in the Pharmaceutical Industry* at 48; CBO, *Increased Competition from Generic Drugs,* at 27-30.

[102] *See* FTC PBM REPORT at 48-54.

[103] *See* Urlep at 34; Lane at 36.

[104] *See* Brugger at 74 (“[W]hat is very important to us to make continued investment in this field is a very clear path towards interchangeability, and what that does is allows companies like ours to innovate in the analytical space and not in the clinical trial space. These clinical trials are a very crude way to detect similarities or differences between these very complex molecules, and the way that we will truly understand these complex macro

molecules in the future is by innovating in this analytical space."); Behrman at 77 ("I couldn't agree ... more that the real advances will come in the analytics and the ability to, to the best of our ability, realize how similar or different these products are and may minimize or shorten or decrease the extent to which certain types of clinical trials are necessary.").

105 Hospira (Wilkie Farr) Comment (12/22/08) at 3 ("If a company pursuing the development of a biosimilar/biogeneric cannot reference any of the innovator's preclinical or clinical data, there would be no incentive to embark on an abbreviated approval pathway.").

106 Hospira Comment (9/30/08) at 1; *but see* BIO Comment (9/30/08) at 3-4 (presumed biogenerics are more expensive to get approved and priced higher than biosimilars).

107 GPhA Comment (9/30/08) at 1-2; *see also* Novartis Comment (9/29/08) at 3 (explaining that interchangeability would "enable direct, head-to-head competition to occur based on price factoring in the front-loaded investment in the research and development of an FOB without the additional cost of a 'back-loaded' investment in the advertising, promotion, and detailing of an FOB. Consequently, competing FOBs that are designated as interchangeable can be anticipated to achieve more rapid and ultimately more substantial market share penetration that those that are not.").

108 *See* Amgen Comment (9/30/08) at 6; CBO S. 1695 Report; Momenta Comment (9/30/08) at 4 ("A designation of "interchangeability" by FDA would significantly increase the competitive impact of a follow-on biologic product and consequently the potential for cost savings."); Novartis Comment (9/29/08) at 9, 11, 24.

109 BIO Comment (12/22/08) at 1; Heldman at 24-28.

110 BIO Comment (12/22/08) at 3; PhRMA Comment (9/30/08) at 1.

111 *See* Charles E. Phelps, *Managing the Market: Regulation and Technical Change in Health Care*, HEALTH ECONOMICS, at 498-546 (3d ed. 2003).

112 *See* Iain M. Cockburn, *The Changing Structure of the Pharmaceutical Industry*, 23 HEALTH AFFAIRS 1:10-22, 14 (2004).

113 Other ways to incentivize innovation include tax credits for R&D costs similar to the tax credits used for orphan drugs. *See* Orphan Drug Act, 21 U.S.C.A. § 360aa-dd (2009). Alternatively, one commenter suggested that a new regulatory scheme be developed to allow for the reporting of R&D costs by pioneer manufacturers and then to have FOB entrants repay a share of these costs. *See* Essential Action Comment (12/22/08) at 4. This system may be difficult to establish and administer because FOBs are similar, not identical to the branded product, and may rely on different FDA findings of safety and effectiveness of the branded product to support regulatory approval.

114 This chapter uses the term "branded exclusivity" rather than "data exclusivity" because current legislative proposals permit an FOB applicant to rely on FDA's finding or conclusion that an approved pioneer drug is safe and effective. This reliance does not involve disclosure to the FOB applicant, or to the public, of the data in the pioneer manufacturers' application. *See* FDA to Petitioners (May 30, 2006) at 6, *available at* http://www.fda.gov/ohrms/dockets/dockets/04P0231/04P-0231-pdn0001.pdf. Further, reliance on the FDA's findings of safety and efficacy of the pioneer biologic provides much less of a benefit in the biologic context than in the small-molecule context, because the FOB will still have substantial R&D expenditures, including clinical testing. *See infra* Ch. 1 at 9, 14-15.

115 Henry Grabowski, *Follow-on Biologics: Data Exclusivity and the Balance Between Innovation and Competition*, 7 NATURE REVIEWS DRUG DISCOVERY 479, 483 (June 2008).

116 American Enterprise Institute ("AEI") Comment (12/10/08) at 5 ("[D]ata exclusivity is a tool that comes into play when patents fail to provide reasonable protection for innovation."); Henry Grabowski *et al., Updating Prior Analyses and Responding to Critiques,* Duke Univ. Dept. Econ. Working Paper, No. 2008-10 (Dec. 22, 2008) at 3 ("[E]xclusivity periods are essential to compensate for some important shortcomings in patent protection for biologics."); Duke University Comment (12/23/08) at 1 ("Data exclusivity periods . . . are an "insurance policy.").

117 F.M. Scherer & David Ross, INDUSTRIAL MARKET STRUCTURE AND ECONOMIC PERFORMANCE, 3d Ed. 621 (1990).

118 The Supreme Court has emphasized the "careful balance" embodied in the patent system: "From their inception, the federal patent laws have embodied a careful balance between the need to promote innovation and the recognition that imitation and refinement through imitation are both necessary to invention itself and the very lifeblood of a competitive economy." Bonito Boats, Inc. v. Thunder Craft Boats, Inc., 489 U.S. 141, 146 (1989).

119 *See* Joseph DiMasi & Cherie Paquette, *The Economics of Follow-on Drug Research and Development*, 22 PHARMACOECONOMICS Supp 2:1-14, 10 (2004). Although this study examined pharmaceutical products primarily, it included several biologic drugs as well.

120 F.M. Scherer, *Markets and Uncertainty in Pharmaceutical Development* 13 (FACULTY RESEARCH WORKING PAPERS SER., HARV. U. JOHN F. KENNEDY SCHOOL OF GOV'T, 2007), *available at* http://ksgnotes1.harvard.edu/Research/wpaper.nsf/rwp/RWP07-039/$File/rwp_07_039_scherer.pdf.

121 *See* F.M. Scherer, *Pharmaceutical Innovation,* (John F. Kennedy School of Government, Harvard Univ., Working Paper No. RWP07-004, 2007) *available at* http://ksgnotes1.harvard.edu/ Research/wpaper.ns

f/rwp/RWP07-004; Scherer, *Markets and Uncertainty,* at 11; *see also* U.S. Cong., Office of Technology Assessment ("OTA"), *Pharmaceutical R&D: Costs, Risks and Rewards*, at 6 (1993), *available at* http://www.princeton.edu/~ota (follow "OTA publications" hyperlink and use search engine there to find article by title) [hereinafter "Pharmaceutical R&D"].

[122] Public policy to increase the U.S. expenditures for research, development and commercialization of federally-funded inventions led to enactment of The University and Small Business Patent Procedures Act of 1980 (also known as "The Bayh-Dole Act"), 35 U.S.C.A. § 200 *et seq*. (2009). This act provides universities and small businesses the right to patent federally funded inventions. Corporations and larger businesses were afforded these same rights pursuant to the Trademark Clarification Act of 1984. 15 U.S.C.A. § 1501 (2009); 35 U.S.C.A. § 210(c) (2009). Privatization of government-funded research was deemed necessary because of a market failure to allocate risk capital to early-stage inventions. Since passage of the Bayh-Dole Act, university-based research has increased by over 800%. *See* Lewis M. Branscomb and Philip Auerwald, *Between Invention and Innovation: An Analysis of Funding for Early Stage Technology Development, prepared for* Nat'l Inst. of Standards and Technology ("NIST"), Dept. of Commerce (2002), *available at* http://www.atp.nist.gov/eao/gcr02-841/contents.htm. While government funds are used on a variety of novel scientific research, corporate funding typically is incremental innovation to support its pre-existing core business, and to "advance its established product and process technologies to better serve existing markets." *Id*. at 4.

[123] Jerry G. Thursby & Marie Thursby, *Enhanced: University Licensing and the Bayh-Dole Act,* SCIENCE, Aug. 22, 2003, at 1052, *available at* http://www.sciencemag.org/cgi/contebt/full/301/5636/1052.

[124] Steven Kozlowski, *Protein Therapeutics and the Regulation of Quality: A Brief History,* BIOPHARM INTERNATIONAL (Oct. 1, 2007), *available at* http://biopharminternational.findpharma.com/biopharm/article /articleDetail.jsp?id=462759&sk=&date=&p ageID=2. In addition to these two classes of therapeutic drug treatments, three additional classes of biotechnological products include: (1) vaccines, which typically are preventative treatments but are under investigation for use as therapeutic treatments; (2) cell therapies and (3) gene therapies which are in clinical development.

[125] *See* NIST, DEP'T. OF COMMERCE, BEYOND MEASURE: A PROFILE OF ATP HEALTH CARE INVESTMENTS (2003), *available at* http://www.atp.nist.gov/atp/brochures/healthcare.pdf (estimating that federal funding accounts for 21 to 25 percent of funding during the "valley of death" period between basic research and product development).

[126] *See* Branscomb, *Between Invention and Innovation,* at Figure 2, p. 33, *see also* Tanuja V. Garde, *Supporting Innovation in Targeted Treatments: Licenses of Right to NIH-funded Research Tools,* 11 MICH. TELECOMM. TECH. L. REV. 249, 277 (2005), *available at* http://www.mttlr.org/voleleven/garde.pdf.

[127] *See* Joseph DiMasi *et al., The Price of Innovation: New Estimates of Drug Development Costs*, 22 J. Health Econ. 151, 162, 166 (2003) (estimating that average time from synthesis of a compound to initial human testing is 52.0 months, from the start of clinical testing to marketing approval is 90.3 months, and the total time is approximately 142 months [11 years, 10 months] for small-molecule drugs); *see also* Grabowski, NATURE, at 481.

[128] DiMasi, *The Economics of Follow-on Drug Research and Development*, at 10.

[129] Scherer, *Markets and Uncertainty,* at 13.

[130] DiMasi, *The Economics of Follow-on Drug Research and Development,* at 9.

[131] *Id;* Scherer, *Markets and Uncertainty,* at 13.

[132] *See e.g*., AEI Comment (12/10/08) at 2 (post approval development of novel biologics continues as science evolves); *see also* John Calfee, *The Golden Age of Medical Innovation,* THE AMERICAN (March/April 2007), *available at* http://www.american.com/archive/2007/march-april-magazine-contents/the-golden-age-of-medical-innovation.

[133] *See, e.g., Patent Reform: The Future of American Innovation, Hearing Before the S. Comm. on the Judiciary,* 110th Cong. (2007) (statement of Kathryn L. Biberstein, Senior Vice President, Alkermes, on behalf of the Biotechnology Industry Organization), *available at* http://judiciary.senate.gov/hearings/hearing.cfm?id=2803.

[134] F.M. Scherer & David Ross, INDUSTRIAL MARKET STRUCTURE AND ECONOMIC PERFORMANCE, 621 (3d ed. 1990); *see also* Amgen, Inc. v. F. Hoffmann-La Roche Ltd., 2008 U.S. Dist. LEXIS 77343 at *169-73 (D. Mass. Oct. 2, 2008 ("Of course, the public derives significant benefits from the innovation generated by the economic incentives in our patent system.").

[135] *See* FEDERAL TRADE COMM'N, TO PROMOTE INNOVATION: THE PROPER BALANCE OF COMPETITION AND PATENT LAW AND POLICY (2003), CH. 3 AT 1 [HEREINAFTER "FTC PATENT REPORT"].

[136] *Id, see also* Arti K. Rai, *Knowledge Commons: The Cost of the Biopharmaceutical Industry*, FIRST MONDAY (June 2007) ("Small biotechnology firms rely on patents, often on technology that is far removed from an end product, for purposes of deterring misappropriation when they market their technology. Patents also help small biotechnology firms negotiate vertical R&D alliances with pharmaceutical firms. For their part, pharmaceutical firms rely on patents on end product drugs for purposes of recouping research and development costs. [footnotes omitted.]"), *available at* http://firstmonday.org/htbin/cgiwrap/bin /ojs/index.php/fm/article/view/1909/1791.

[137] *Id.* at 1-2.

[138] *See* BIO Comment (9/30/08) at 10-12.

[139] BIO Comment (9/30/08) at 12; *see also* Bruce S. Manheim *et al., 'Follow-On Biologics': Ensuring Continued Innovation In The Biotechnology Industry*, 25 HEALTH AFFAIRS 2: 394-404, 397(March/April 2006) ("the identity of a [biologic] product is clearly dependent upon the process used to manufacture the product.").

[140] One commenter noted that not all biologic manufacturing processes are patented, and may restrain entry by FOBs because they are trade secrets in the possession of the branded manufacturers. Essential Action Comment (12/22/08) at 3, and fn. 3 (quoting Gregory Mandel, *The Generics Biologics Debate: Industry's Unintended Admission That Biotech Patents Fail Enablement*, 11 Va. J.L. & Tech. at 66 (2006)). Trade secrets often cover methods of making biologic products. *See e.g.,* Letter from Director Steven K. Glason, FDA Center for Evaluation and Research ("CDER") to Petitioners (May 30, 2006) at 9, *available at* http://www.fda.gov/ohrms/dockets/dockets/04P0231/04P-0231-pdn0001.pdf [hereinafter "FDA's Second Response to Omnitrope CPs"]; Reinhardt Comment (10/19/08) at 2; Novartis Comment (9/29/08) at 2; Momenta Comment (9/30/08) at 40.

[141] FTC PATENT REPORT, at Ch. 3 at 1; Michael Heller *et al., Can Patents Deter Innovation? The Anticommons in Biomedical Research*, 280 SCIENCE at 698-701 (MAY 1, 1998) (posits the concern that while biomedical patents spur private investment to support innovation but risks paradoxically the proliferation of fragmented and overlapping intellectual property rights which may restrict the number of commercialized products for improving human health); Suzanne Scotchmer, *Standing On The Shoulders Of Giants: Cumulative Research And Patent Law,* JOURNAL OF ECONOMIC PERSPECTIVES, 5 AMERICAN ECONOMIC ASSOCIATION 1: 29-41 (1991); *see also* HHS, NIH, *Principles and Guidelines for Sharing of Biomedical Resources,* 64 Fed. Reg. No. 246, 72090 (December 23,1999) (ensuring that the conditions imposed on the transfer of research tools will facilitate further biomedical research, consistent with the requirements of the Bayh-Dole Act).

[142] John Walsh *et al., Effects of Research Tools Patents and Licensing on Biomedical Innovation,* PATENTS IN THE KNOWLEDGE BASED ECONOMY, 285-336 (2003); John Walsh *et al., View From the Bench: Patents and Material Transfers*, 309 SCIENCE 2003-03 (September 23, 2005); Office of Industries, U.S. International Trade Comm'n, *Patenting Trends and Innovation in Industrial Biotechnology* (October 2008).

[143] *See. e.g.*, AEI Comment (12/10/08) at 5 ("[D]ata exclusivity is a tool that comes into play when patents fail to provide reasonable protection for innovation."); BIO Comment (9/30/08) at 4; Duke University Comment (12/22/08) at 3.

[144] Norman at 156.

[145] Kushan at 180.

[146] Manspeizer at 148-49; BIO Comment (12/22/08) at 5; *see also* Wyeth Comment (12/18/08) at 8..

[147] *See e.g.,* Wyeth Comment (12/18/08) at 8; Manheim, *Follow-On Biologics.*

[148] Seide at 150-52; BIO Comment (12/22/08) at 5-6 (citing case law).

[149] Kepplinger at 158-59 (further explaining that under the PTO's current practice for a molecule patent claim to a method of use or function, the PTO will restrict the function claim to the narrow molecule and not broaden it across variations of that molecule.)

[150] Dow at 166-67.

[151] Kepplinger at 157-58*; see generally* Patent and Trademark Office, Written Description Training Materials (2008), *available at* http://www.uspto.gov/web/menu/written.pdf.

[152] *See* Goldman at 164-65, Kepplinger at 158-59; Manspeizer at 148-49.

[153] Dow at 166-67, 169; Manspeizer at 148-49.

[154] Hospira (Wilkie Farr) Comment (12/22/08) at 2 ("Biologics are large molecules, and product patents typically only claim their 'active' regions. [footnote omitted] These active regions engage the molecule with its surrounding environment and create the therapeutic effect. Thus, while biosimilars might be similar, but not identical, their functionality will likely require resolution of product claims covering the biologic's active region, regions that will often be shared by both the reference biologic and the biosimilar."); Leicher at 161-62; Winston & Strawn ("W&S") Comment (12/22/08) at 4.

[155] Pearce at 169.

[156] Hospira (Wilkie Farr) Comment (12/22/08) at 3; Amgen, Inc. v. F. Hoffmann-La Roche Ltd.., 2008 U.S. Dist. LEXIS 77343 at * 13 (D. Mass. Oct. 2, 2008); Genzyme Corp. v. Transkaryotic Therapies, Inc., 346 F.3d 1094 (Fed. Cir. 2003).

[157] Hospira (Wilkie Farr) Comment (12/22/08) at 2.

[158] Pearce at 144-45.

[159] Momenta Comment (12/22/08) at 8.

[160] Panelists explained the number of patents per biologic product, including platform patents, is substantial, resulting in significant "stacking" of patents (or royalties) compared to the small-molecule patent estates. *See* Dow at 185; Sauer at 261; Seide at 238; Duncan Bucknell Co. Comment (1/9/09) at 9; Momenta Comment (9/30/08) at 6-7; Wyeth Comment (9/30/08) at 4.

[161] Essential Action Comment (12/22/08) at 3, and fn. 3 (quoting Gregory Mandel, *The Generics Biologics Debate: Industry's Unintended Admission That Biotech Patents Fail Enablement*, 11 Va. J.L. & Tech. at 66 (2006)); W&S Comment (12/22/08) at 5;.

[162] W&S Comment (12/22/08) at 3.

[163] Leicher at 162-63; Dow at 185.

[164] Hospira (Wilkie Farr) Comment (12/22/08) at 3. Submarine patents result from older patent applications that are not published. Because the applications are not made available 18 months after filing, competitors cannot use the applications to determine whether their FOB products in R&D are likely to infringe potential issued patents. This also creates uncertainty for competitors.

[165] F.M. Scherer & David Ross, INDUSTRIAL MARKET STRUCTURE AND ECONOMIC PERFORMANCE, at 621 (3rd Ed., 1990); *see also Patent Reform: The Future of American Innovation, Hearing Before the S. Comm. on the Judiciary,* 110th Cong. (2007) (statement of Kathryn L. Biberstein, Senior Vice President, Alkermes) at 2; *Patent Reform Act of 2007: Hearing on H.R. 1908 Before the H. Subcomm. on Courts, the Internet, and Intellectual Property of the H. Comm. on Judiciary,* 110th Cong. 65 (2007) (statement of Kevin Sharer, CEO of Amgen), *available at* http://judiciary.house.gov/hearings/April2007/Sharer070426.pdf., *Stifling or Stimulating - The Role of Gene Patents in Research and Genetic Testing Before the H. Comm. on the Judiciary, Subcomm. on Courts, the Internet, and Intellectual Property*, 110th Cong. (2007) (statement of Jeffrey P. Kushan on Behalf of BIO), *available at* http://judiciary.house.gov/hearings/pdf/Kushan071030.pdf.

[166] *Id., see also* Dow at 185; Sauer at 261; Seide at 238; Duncan Bucknell Co. Comment (1/9/09) at 9; Momenta Comment (9/30/08) at 6-7; Wyeth Comment (9/30/08) at 4.

[167] Bio-Technology Gen. Corp. v. Genentech, Inc., 267 F.3d 1325, 1333 (Fed. Cir. 2001) (reversing trial court's finding of invalidity for lack of enablement regarding method of production patent); *see also* Press Release, Genentech, Genentech Receives Final Notification Upholding Cabilly Patent in Reexamination Proceeding (Feb. 24, 2009), *available at* www.gene.com (announcing PTO issuance of Notice of Intent to Issue a Reexamination Certificate confirming patentability of all claims of the Cabilly Patent (U.S. Pat. No. 6,331,415) claiming methods of making recombinant cells expressing both an immunoglobulin light chain and heavy chain used in genetically-engineered monoclonal antibodies).

[168] John R. Thomas, *Toward a Theory of Marketing Exclusivities* at 32-33 (2009) (forthcoming) ("Biotechnology products may commonly be defined through multiple techniques, including their structure, chemical or physical characteristics, and method of preparation, that in combination are capable of providing a potent shield against would-be competitors.").

[169] Christopher M. Holman, *Is Lilly Written Description a Paper Tiger?: A Comprehensive Assessment of the Impact of Eli Lilly and Its Progeny in the Courts and PTO*, 17 ALB. L.J. SCI. & TECH. 1, 44 (2007).

[170] PTO, WRITTEN DESCRIPTION TRAINING MATERIALS (2008), *available at* http://www.uspto.gov/ web/ menu/written.pdf.

[171] *See, e.g.*, Infigen, Inc. v. Advanced Cell Technology, Inc., 65 F. Supp.2d 967,975 (W.D. Wis. 1999) ("It is black letter law that claims are not limited to the embodiment described in the patent specifications. Moreover, a patent claim may encompass uses not anticipated by the inventor and therefore not described in the patent.") (citations omitted). This principle extends beyond percent identity claims. Capon v. Eshhar, 418 F.3d. 1349, 1359 (Fed. Cir. 2005) (reversing BPAI interpretation of written description and holding, "It is not necessary that every permutation within a generally operable invention be effective in order for an inventor to obtain a generic claim, provided that the effect is sufficiently demonstrated to characterize a generic invention."); Invitrogen Corp. v. Clontech Labs., 429 F. 3d 1052, 1071-74 (Fed. Cir. 2005) (holding that disclosure of a known protein variant satisfied written description for claims encompassing engineered protein variants with shared function);*id* at 1073 ("Enablement does not require the inventor to foresee every means of implementing an invention at pains of losing his patent franchise."), *see generally* Christopher M. Holman, *Is Lilly Written Description a Paper Tiger?: A Comprehensive Assessment of the Impact of Eli Lilly and Its Progeny in the Courts and PTO*, 17 ALB. L.J. SCI. & TECH. 1 (2007) (a comprehensive review of federal court and PTO Board of Patent Appeals and Interferences decisions reveals no support for the proposition that since the Lilly decision, which purportedly tightened the written description requirements for biotechnology drug molecule claims, patentees have not been able to obtain patents with sufficiently broad scope.).

[172] Christopher M. Holman, *Is Lilly Written Description a Paper Tiger?: A Comprehensive Assessment of the Impact of Eli Lilly and Its Progeny in the Courts and PTO*, 17 ALB. L.J. SCI. & TECH. 1, 47 (2007).

[173] Biogen, Inc. v. Berlex Labs, 318 F.3d 1132, 1140-42 (Fed. Cir. 2003) (affirming trial court's finding that there was no literal infringement and vacating summary judgment of non-infringement under the doctrine of equivalents); Genentech, Inc. v. The Wellcome Found. Ltd., 29 F.3d 1555, 1569 (Fed. Cir. 1997) (reversing trial court's denial of defendant's JMOL in part because an element of the doctrine of equivalents was not met); Novo Nordisk of N. America, Inc., v. Genentech, Inc., 77 F.3d 1364, 1367-71 (Fed. Cir. 1996) (vacating preliminary injunction on grounds that district court erred in finding literal infringement where "direct expression" of human growth hormone did not cover alleged infringers "cleavable fusion" process for producing the hormone); Hormone Research Foundation, Inc., v. Genentech, Inc., 904 F.2d 1558, 1563-67 (Fed. Cir. 1990) (finding no literal infringement, vacating infringement under the doctrine of equivalents,

holding that the patentee did not intend "corresponding to" and "similar" to have the same meaning, and ruling that "corresponding to" reflected true identity).

[174] Genentech, Inc. v. Chiron Corp., 112 F.3d 495, 501 (Fed. Cir. 1997) (reversing district court's claim interpretation in interference proceeding because it was "not the broadest, reasonable interpretation of the count.") (citing Genentech, Inc. v. The Wellcome Found., Ltd., 29 F.3d 1555 (Fed Cir. 1997)); Amgen, Inc. v. Chugai Pharm. Co., 927 F.2d 1200, 1219 (Fed. Cir. 1991) (affirming trial court's finding that certain claims covering purified and isolated DNA sequences encoding EPO and host cells transformed or transfected with a DNA sequence were valid and infringed and reversing finding that other claims were enabled); Amgen, Inc. v. Hoffmann-LaRoche Ltd., 2008 U.S. Dist. LEXIS 77343 (D. Mass. 2008) (granting permanent injunction for infringement of claims directed to a specific amino acid sequence); *see also* Amgen Inc. v. Hoechst Marion Roussel, 579 F. Supp. 2d 199, 210 (D. Mass. 2008); Chiron v. Genentech, Inc., 268 F. Supp. 2d 1126, 1138 (E.D. Cal. 2008); Genentech, Inc. v. Insmed Inc., 436 F. Supp. 2d 1080, 1091-92 (N.D. Cal. 2006) (granting patent holder's partial motion for summary judgment and finding literal infringement of a patent claiming "Preparation of Human IGF via Recombinant DNA I Technology").

[175] Amgen, Inc. v. F. Hoffmann La Roche, Ltd., 2008 U.S. Dist. LEXIS 77343 (D. Mass. Oct. 2, 2008).

[176] *See e.g.*, Bayer HealthCare LLC v. Abbott Laboratories, Civ. Action No. 6:08cv507 (E.D. Tex. Dec. 24, 2008), complaint *available at* http://www.patentbaristas.com/wp/wp-content/uploads/2009/01/bayer.pdf. Abbott countersued in Massachusetts: Abbott Laboratories v. Bayer Healthcare LLC, 09cv40002, U.S. District Court for the District of Massachusetts, 09cv40002, (D. Mass. June 5, 2008).

[177] BIO Comment (9/30/08) at 22 (citing industry experience over more than two decades of biotechnology patent litigation).

[178] FTC PATENT REPORT, Ch. 3 at 1-2.

[179] Phillips at 100-01 ("[I]f there is no chance to recoup the capital outlay, then the investment won't be made."). This panelist also suggested that there is a dynamic effect to a branded exclusivity period in that it "is going to change the status quo for investment decisions made by innovator companies;" *see also* AEI Comment (12/10/08); Eli Lilly Comment (12/19/08) at 2; Wyeth Comment (12/18/08) at 2 ("Just as certainty spurs innovation and advances that benefit patients, lack of certainty in the pharmaceutical and biotechnology industries hinders innovation").

[180] Johnson & Johnson Comment (3/17/09) at 6.

[181] Grabowski, *Follow-on Biologic.*

[182] Duke University Comment (12/22/08); PhRMA Comment (12/22/08).

[183] *See, e.g.*, BIO Comment (9/30/08) at 17 ("Failure to provide substantial data exclusivity would fundamentally alter the ability of biotechnology companies to continue to innovate because these companies, in order to secure the necessary resources from venture capital firms and other funding sources, must have some certainty that the can prevent free-riding on their investment in the development of new breakthrough therapies for a substantial period of time.")

[184] Wyeth Comment (12/18/08) at 10.

[185] *Id.* at 9.

[186] BIO Comment (12/22/08) at 6.

[187] Henry Grabowski *et al.*, *Generic Competition and Market Exclusivity Periods in Pharmaceuticals*, MANAGERIAL AND DECISION ECONOMICS 28 (2007) 491–502, at 493 (the study examined 251 products that encountered generic entry between 1995 and 2005).

[188] Heldman at 117; *see also* Essential Action Comment (12/22/08) at 2 ("[F]ree from competitive pressures they can set a price that allows them to earn profits, and not just recoup their R&D costs.").

[189] Grabowski at 117-18 ("[P]rice is going to be driven by your interaction with payers and other competitors.").

[190] Zuckerman Spaeder Comment (12/22/08) at 12.

[191] Teva Comment at 5 ("Evergreening will multiply the economic costs of expanding monopoly protection via exclusivity arrangements. Brand companies can, and routinely do, make relatively minor changes to their existing products in order to restart their monopoly-protection clocks.").

[192] Brugger at 74; Behrman at 77-79; Grabowski at 80; Barr Comment (12/19/08) at 1-2 (the anticompetitive barrier to FOB competition is longer than just the term of the exclusivity period, as FOB cannot file its application until the day after the period expires, and entry is further delayed for year(s) while FDA reviews the FOB's application.); Momenta Comment (12/22/08) at 1-3; Novartis Comment (9/29/08) at 10 ("With no market access, there is only limited incentive to make safe and effective competing products . . . expand the market with new indications . . . .implement more efficient and cost-effective manufacturing that potentially can enable reductions in costs of goods.").

[193] Grabowski at 128; Horton at 129-30; Philips at 132-33.

[194] *See* John E. Calfee, *Facing Reality on Follow-On Biologics*, AEI (2007), *available at* http://www.aei.org/docLib/20070423_200704AHPOg.pdf.

[195] AEI Comment (12/10/08) at 2.

[196] Wyeth Comment (12/18/08) at 9.

[197] BIO Comment (12/22/08) at 7-8.

[198] Brill at 133-34.

[199] *See generally* DiMasi, *The Economics of Follow-on Drug Research and Development*; Ian Cockburn, *The Changing Structure of the Pharmaceutical Industry*, 23 HEALTH AFF. 1:10-22 (2004).

[200] William J. Bauomol, THE FREE-MARKET INNOVATION MACHINE: ANALYZING THE GROWTH MIRACLE OF CAPITALISM (2002) at 33 (capitalism benefits society not just through price competition but also through systematic innovation races among all firms in an innovating industry as they vie for consumers and dare not fall behind the others in new products and processes).

[201] See Abbott Labs. v. Teva Pharm. USA, Inc., 432 F. Supp. 2d 408, 413 n.1 (D. Del. 2006) for an example of litigation alleging this type of strategy. *See also* Herbert Hovencamp, Mark D. Janis & Mark A. Lemley, IP AND ANTITRUST, § 12.5 (2006).

[202] *See generally* DiMasi, *The Economics of Follow-on Drug Research and Development.* Although this study examined pharmaceutical products primarily, it included several biologic drugs as well.

[203] Scherer, *Markets and Uncertainty*, at 13.

[204] Grabowski, *Generic Competition and Market Exclusivity Periods in Pharmaceuticals,* at 493.

[205] Charles Clift, *The Value of Patent Term Extensions to the Pharmaceutical Industry in the USA,* 5 J. GEN. MED. 201-208 (Apr. 2008).

[206] Barr Comment (9/30/08) at 12.; PhRMA Comment (9/30/08) at 19 ("As in the case of generic drugs, any regulatory approval pathway for FOBs would involve complex scientific and legal considerations that can and should be raised through appropriate mechanisms, such as citizen petitions. Innovator companies have extensive knowledge about their products, and are often in the best position to bring to FDA's attention complex regulatory and scientific issues regarding appropriate approval standards.").

[207] *See, e.g.,* H. 1548, 111th Cong. § 101(k)(2)(B)(iii) (2009) (guidelines required for assessing immunogenicity); *id.* at § 101(k)(4)(B) (guidelines required for assessing interchangeability); and *id.* at § 101(k)(5)(C) (guidelines required for Risk Evaluation and Mitigations Strategy).

[208] *See* FEDERAL TRADE COMM'N, GENERIC DRUG ENTRY PRIOR TO PATENT EXPIRATION: AN FTC STUDY (JULY 2002), at iiii, *available at* http://www.ftc.gov/os/2002/07/genericdrugstudy.pdf.

[209] *See infra* Chapter 3.

[210] *See* BIO Comment (5/1/09) at 7-9 (Benjamin Roin, *Unpatentable Drugs and the Standards of Patentability* 87 TEX. L. REV. (forthcoming)).

[211] *See* Appendix B for a description of the marketing exclusivities for small-molecule drug products.

[212] *See* Orphan Drug Act ("ODA"), 21 U.S.C.A. § 360aa *et seq.* (2009), 21 C.F.R. § 316 *et seq.;* FDA, Office of Orphan Products Dev't, Cong. Findings For the ODA (" [B]ecause so few individuals are affected by any one rare disease or condition, a pharmaceutical company which develops an orphan drug may reasonably expect the drug to generate relatively small sales in comparison to the cost of developing the drug and consequently to incur a financial loss; there is reason to believe that some promising orphan drugs will not be developed unless changes are made in the applicable Federal laws to reduce the costs of developing such drugs and to provide financial incentives to develop such drugs.") *available at* http://www.fda.gov/orphan/oda.htm. It is likely that the patents for orphan drugs and not the 7-year ODA exclusivity period provide the greatest incentive to innovators. *See* Robert Rogoyski, *The Orphan Drug Act and the Myth of the Exclusivity Incentive,* 7 COLUM. SCI. & TECH. L. REV 2 (2006), http://www.stlr.org/volumes/volume-vii-2005-2006/rogoyski/. According to one study, the majority of orphan drugs are protected by patents with both a broader scope than the disorder specific ODA, and a longer duration than the 7-year ODA exclusivity period. *Id.* at 18, Figure 1.

[213] Such an approach would require, at a minimum, R&D cost information to which the FTC does not have access.

[214] The cost of capital is the annual rate of return that an investor would require. Grabowski, NATURE at 480.

[215] *See Patent Reform: The Future of American Innovation, Hearing Before S. Judiciary Comm.*, 110th Cong. (2007) (statement of Kathryn L. Bieberstein on behalf of the Biotechnology Organization) ("The biotechnology industry, fueled by the strength of the U.S. patent system, has provided jobs for over 200,000 people in the United States, and has generated hundreds of drug products, medical diagnostic tests, biotech crops, and environmental products."), *available at* http://judiciary.senate.gov/hearings/testimony.cfm?id=2803&wit_id=6508 [hereinafter, "Bieberstein Statement"].

[216] F.M. Scherer & David Ross, INDUSTRIAL MARKET STRUCTURE AND ECONOMIC PERFORMANCE, 621 (3d ed. 1990).

[217] BIO Comment (9/30/08) at 22 ("Biotechnology patent disputes today can be adjudicated within a relatively stable doctrinal framework that is expected to solidify further as biotechnology matures both as a science as an industry.").

[218] To the extent that a pre-approval process is designed to address allegations of poor biotechnology patent quality, these issues may be better addressed in efforts to examine patent reform more broadly. Recent Congressional testimony addressed biotechnology issues in the context of patent reform. *See, e.g., Patent Reform in the 111th Congress: Legislation and Recent Court Decisions: Hearings Before the S. Judiciary Comm.*, 111th Cong. (2009).

[219] 35 U.S.C.A. § 271(e)(1) (2009) ("It shall not be an act of infringement to make, use, offer to sell, or sell within the United States or import into the United States a patented invention (other than a new animal drug or

veterinary biological product (as those terms are used in the Federal Food, Drug, and Cosmetic Act and the Act of March 4, 1913) which is primarily manufactured using recombinant DNA, recombinant RNA, hybridoma technology, or other processes involving site specific genetic manipulation techniques) solely for uses reasonably related to the development and submission of information under a Federal law which regulates the manufacture, use, or sale of drugs or veterinary biological products.").

220 Merck KGaA v. Integra Lifesciences I, Ltd., 545 U.S. 193 (2005).

221 See Appendix B for a description of the patent resolution process in the Hatch-Waxman Act.

222 This "30-month stay" expires at the earliest of: (1) the date the patent(s) expire; (2) a final determination of non-infringement or patent invalidity by a court in the patent litigation; or (3) the expiration of the 30 months from receipt of notice of the paragraph IV certification.

223 *See, e.g.*, Amgen (9/30/08) Comment at 20 (certainty may increase FOB uptake if physicians are more likely to prescribe an FOB after the product has cleared patent hurdles); Essential Action Comment (12/20/08) at 7 ("should be to clear patent claims so that a) invalid patents do not delay investment in, or introduction of, generic or similar products; b) non-applicable patents do not delay investment in, or introduction of, generic or similar products; and c) all potential patent claims are resolved in advance of any applicable marketing exclusivities."); Manspeizer at 229; Leicher at 232; and Dow at 295-96.

224 Hospira Comment (9/30/08) at 7 ("Due to the greater uncertainty surrounding the valid scope of patents and the lack of jurisprudence resulting from an immature biopharmaceutical industry as compared to a small molecule drug . . . this will operate as a significant disincentive to launch of a biogeneric and will thus operate as a disincentive to competition."); Seide at 238; and Siwik at 224-25.

225 Amgen Comment (9/30/08) at 20; BIO Comment (9/30/08) at 20.

226 Leicher at 232.

227 Sauer at 227.

228 *Id.* at 271.

229 Wyeth Comment (9/30/08) at 9; *see also* Wyeth Comment (12/18/08) at 13 ("In order to provide certainty to all parties concerning the outcome of any patent resolution mechanism, a linkage system is required.").

230 Novartis Comment (9/29/08) at 19; and Goldman at 231 ("Decoupling will avoid premature challenges to biotech patent estates ahead of the prospect of imminent commercialization, and current law provides robust protection for those rights when infringement occurs.").

231 Goldman at 230 ("There's not a single product that hasn't come on the market in which launching at risk hasn't been a key issue. And companies are – all of us here have the ability to take that business risk into consideration and decide whether or not to launch at risk."); *see also id.* ("[T]he need for an early resolution, early litigation because of the fear of launching at risk is not a serious one we contend."); and Novartis Comment (9/29/08) at 19.

232 Goldman at 230-31 (explaining that "[t]here's no artificial act of infringement in the European scheme as well, so it's a real aberration.").

233 Siwik at 273-74.

234 Kushan at 276-77.

235 Siwik at 224-25 ("[I]t's important to have a mechanism in the bill for resolving certain patent disputes concurrent with FDA review, but the big but is, if the system doesn't work, if whatever this patent mechanism is doesn't work, I guess work in the sense that it can delay market launch.").

236 Goldman at 231 ("You litigate one patent followed by another patent, and that can really extend the litigation pre-approval.").

237 Kepplinger at 267 ("[I]t seems like one of the lessons from Hatch-Waxman, and many people have talked about it, is that there's quite a lot of litigation, and it seems like in designing the situation, we should be looking to try to reduce the litigation because it is just a lot of money that could probably be better spent on other things, like designing more pharmaceuticals.").

238 Goldman at 242-43; Novartis Comment (9/29/08) at 18-19.

239 Goldman at 231 ("[A]nd besides that, we also see in those cases that there's serial litigation. You litigate one patent followed by another patent, and that can really extend the litigation pre-approval. Post approval, there's no incentive for serial litigation. You would want to bring your best patents quickly to get the product off the market.").

240 BIO Comment (9/30/08) at 21 ("Patent litigation would be premature if it were allowed to commence before a determination that the FOB application in question is complete and in condition for review without additional clinical studies."); *see also* Wyeth Comment (12/18/2008) at 13 ("[a] patent resolution proceeding should not be initiated at a point in time that is too early, when the details of the biosimilar product are not yet fully defined or manufacturing processes still are subject to change.").

241 Goldman at 240-41 ("[I]t surprises me that . . . the companies that are worried about not having enough money are the ones that are advocating jumping into expensive litigation 30 months early. I would think that you would want to avoid that, the litigation . . . . [Y]ou may in fact be bringing on expensive litigation costs earlier when you might not want to do that."); *see also* Siwik at 225.

[242] *See, e.g.*, David Reiffen and M.R. Ward, *"Branded Generics" As A Strategy To Limit Cannibalization of Pharmaceutical Markets,* 28 MANAGERIAL AND DECISION ECONOMICS 251, 264 (2005), *available at* http://ftc.gov/be/healthcare/wp/12_Reiffen_BrandedGenericsAsAStrategy.pdf

[243] BIO Comment (9/30/08) at 22.

[244] MedImmune Inc. v. Genentech, Inc. 549 U.S. 118, 137 (2007) (holding that a licensee has standing to bring a declaratory judgment action for non-infringement); Merck KGaA v. Integra Lifesciences I, Ltd., 545 U.S. 193, 206-08 (2005) (clarifying 35 U.S.C.A. 271(e)(1)).

[245] *See, e.g.,* Amgen, Inc. v. Chugai Pharm. Co., 927 F.2d 1200, 1219 (Fed. Cir. 1991) (affirming trial court's finding that certain claims covering purified and isolated DNA sequences encoding EPO and host cells transformed or transfected with a DNA sequence were valid and infringed and reversing finding that other claims were enabled); Amgen, Inc. v. Hoffmann-LaRoche Ltd., 2008 U.S. Dist. LEXIS 77343 (D. Mass. 2008) (granting permanent injunction for infringement of claims directed to a specific amino acid sequence); *see also,* Amgen Inc. v. Hoechst Marion Roussel, 579 F. Supp. 2d 199, 210 (D. Mass. 2008); Bayer HealthCare LLC v. Abbott Labs., Civ. Action No. 6:08cv507 (E.D. Tex. Dec. 24, 2008). Abbott countersued in Massachusetts: Abbott Labs. v. Bayer Healthcare LLC, 09cv40002, U.S. District Court for the District of Massachusetts, 09cv40002 (D. Mass. June 5, 2009).

[246] Prior to 2003, if a branded manufacturer listed an additional patent in the Orange Book *after* the generic applicant filed its ANDA, more than one 30-month stay could be generated. The generic applicant was required to re-certify to this later-listed patent, and if, upon notice of the generic's re-certification, the brand-name company sued within 45 days, then FDA approval of the generic's previously filed ANDA was stayed for *an additional* 30-months from the notice date or until a court decision in the newly instituted patent litigation. FTC GENERIC DRUG STUDY at iii. In 2003, Congress amended the Hatch-Waxman Act to address this problem, "allowing lower-priced generic products to enter the market more quickly." Joint Explanatory Text to the MMA Conference Agreement, H.R. Conf. Rep. No. 108-391, at 836 (2003), *reprinted in* 2004 U.S.C.C.A.N. 2187. Now, a generic applicant who amends a pending ANDA to include Paragraph IV certifications to later-listed patents is not subject to a 30-month stay on the amended certification. 21 C.F.R. § 314.94(a)(l2)(vi). This conduct is not unexpected. *See* Robert H. Bork, THE ANTITRUST PARADOX 347 (1978) ("The modern profusion of [. . .] governmental authorities offers almost limitless possibilities for abuse.").

[247] *See, e.g.*, Teva Pharm., USA, Inc. v. FDA, 182 F.3d 1003 (D. C. Cir 1999), *see also* Siwik at 289 ("[I]n Hatch-Waxman we learned that there are rules, but if there are no sticks, the rules are going to go out the window.").

[248] To the extent that the branded company brings suit in a slow-moving venue, the FOB has a variety of tools to force expeditious resolution of its case.

[249] Norman at 201.

[250] Wyeth Comment (12/18/2008). Some panelists explained that there should be "[f]ull disclosure by all participants early in the patent resolution mechanism," calling for patent holders to provide "full disclosure of the patents at issue in any dispute" while FOBs would provide "full disclosure of their application for regulatory approval, including all manufacturing process details." *Id.*

[251] Patent applications filed on or after November 29, 2000 are published eighteen months after the effective filing date of the application. One commenter noted the existence of "submarine" patent applications, a subset of patent applications filed before November 29, 2000 that are not published until the patent is granted. Applications with an effective filing date on or after June 8, 1995 expire 20 years from filing. Applications with an effective filing date before June 8, 1995 expire 17 years from patent grant. *See* Hospira Comment (9/22/08) at 4-5. While this can present issues of extended patent terms for old technology, this problem applies across the industry and likely does not outweigh the likely anticompetitive effects of a notice provision.

[252] *See* http://www.uspto.gov/main/patents.htm.

[253] 35 U.S.C.A. § 287(a) (2009).

[254] American Medical Sys., Inc. v. Medical Eng'g Corp., 6 F.3d 1523, 1538 (Fed. Cir. 1993).

[255] Amstead Indus., Inc. v. Buckeye Steel Castings Co., 24 F.3d 178, 185-87 (Fed. Cir. 1994) (finding lack of notice where the letter did not specifically charge the recipient with infringement and did not identify an infringing device).

[256] 35 U.S.C.A. § 287(b)(4) (2009).

[257] See 21 U.S.C.A. § 355(b)(1) (2009) and Appendix B for a detailed description of the Hatch-Waxman abbreviated drug approval process.

[258] 21 U.S.C.A. §355(j)(2)(A)(vii)(IV) and § 355(j)(5)(B)(iii) (2009).

[259] Fed. Trade Comm'n, GENERIC DRUG ENTRY PRIOR TO PATENT EXPIRATION: AN FTC STUDY 18 (July 2002), *available at* http://www.ftc.gov/os/2002/07/genericdrugstudy.pdf.

[260] Seide at 266.

[261] Kushan at 257-58 ("[T]he notice should include a detailed description of the FOB's product, including the amino acid sequence produced, the nucleic acid sequence, expression technologies, process technologies, manufacturing process information, molecular structure, formulation, patent certifications, molecular identity and intended uses.").

[262] Fed. Trade Comm'n and U.S. Dep't of Justice, ANTITRUST GUIDELINES FOR COLLABORATIONS AMONG COMPETITORS 12 (2000).

[263] Patent law places the burden of proof of demonstrating infringement on the patent holder. The patent holder may not need the FOB's application to establish infringement if the FOB's product already is approved. In addition, a notice provision is likely to have the effect of shifting the burden of proof such that the FOB has to demonstrate that it does not infringe the patent, rather than under patent law having the patent holder show that its patent has been infringed.

[264] Of course, these arguments apply with equal force in the opposite scenario if the FOB manufacturer were to obtain confidential information regarding the branded product.

[265] The pioneer manufacturer-owned patents may be out-licensed further to additional third parties.

[266] Kushan at 237.

[267] Seide at 238 ("The technology platform patents are very important . . . and so there has to be some way of resolving third-party patents as well if they're known.").

[268] Siwik at 226.

[269] Barr Comment (9/30/2008) at 10; CCPM Comment (9/30/2008) at 8; GPHA Comment (9/30/08) at 6; Teva Comment (9/30/08) at 6; Leicher at 260 ("If you limit it to the key patents that are built around the product that the brand company controls, I think you've got it simplified.").

[270] Schultz at 290; Essential Action Comment (12/20/2008) at 6 ("[I]nitial registrants should be required at the time of the application to indicate any granted or filed patents that they believe apply to the biologic for which they seek marketing approval. This should include both patents granted to the registrant or which have been licensed to them. They should be required to update this list for any new patent filings, within a statutorily defined period, perhaps 30 days. Failure to disclose should forfeit the right to enforce.").

[271] Siwik at 289 ("If the overall scheme is fair and balanced, maybe we don't need to worry about huge sticks to make people participate, but in Hatch-Waxman we learned that there are rules, but if there are no sticks, the rules are going to go out the window. There were statutory definitions of what patents could go in the Orange Book, and there were a few companies that abused that, and a list of other patents triggered a lot of 30 month stays, and a lot of litigation delays, but no penalties for doing it.").

[272] Goldman at 287; Kushan at 293 (current laws exists to manage parties who timely fail to enforce their patent rights); PhRMA Comment (12/22/08) at 5.

[273] PhRMA Comment (12/22/08) at 5 (The proposal also "could create artificial incentives to litigate, which would waste time and money and impose burdens that would not be beneficial for the patent or the judicial system.").

[274] Seide at 288.

[275] *See, e.g.*, Sandoz, Inc. v. F.D.A., 439 F. Supp. 2d 26 (D.C. Cir. 2006) (denying generic competitor's motion for injunctive relief against FDA re-listing brand patents); Purepac Pharm. Co. v. TorPharm, Inc., 354 F.3d 877, 886-88 (D.C. Cir. 2004) (upholding FDA decision to delist a patent incorrectly listed for the wrong drug).Dr. Reddy's Labs., Inc. v. Thompson, 302 F. Supp. 2d 340, 355 (D.N.J. 2003) (upholding FDA delisting of an expired patent and not to award exclusivity to an ANDA applicant who filed a paragraph IV certification before the patent's expiration);

[276] *See, e.g.*, American Bioscience, Inc. v. Thompson, 269 F.3d 1077, 1080 (D.C. Cir. 2001) (recognizing that the FDA "has refused to become involved in patent listing disputes, accepting at face value the accuracy of NDA holders' patent declarations and following their listing instructions"); Purepac Pharm. Co. v. Thompson, 238 F. Supp. 2d 191, 196 (D. D.C. 2002) ("The duty to ensure that the Orange Book only lists patents that actually claim approved drugs . . . lies with NDA holders.") (citing Watson Pharm., Inc. v. Henney, 194 F. Supp. 2d 442, 445-46 (D. Md. 2001) ("In making its decision to list a patent . . . it is entirely appropriate and reasonable for the FDA to rely on the patentee's declaration as to coverage, and to let the patent infringement issues play out in other, proper arenas, as is the clear intent of the Hatch-Waxman Amendments.")).

[277] For example, if the rules required a pioneer manufacturer to provide a list of patents within 30 days, but one patent is left off and corrected on day 31, does this omission limit enforceability against that FOB manufacturer? There are countless ways in which parties may inadvertently violate the rules for which a "sue-or-lose" provision would extinguish their patent rights.

[278] 21 U.S.C.A. § 355 (j)(5)(B)(iv)(I) (2009).

[279] Mova v. Shalala, 140 F.3d 1060, 1074 (D.C. Cir. 1998).

[280] *See* David Reiffen & Michael Ward, *"Branded Generics" As A Strategy To Limit Cannibalization of Pharmaceutical Markets,* 28 MANAGERIAL AND DECISION ECONOMICS, 251, 264 (2005), *available at* http://ftc.gov/be/healthcare/wp/12_Reiffen_BrandedGenericsAsAStrategy.pdf.

[281] *See* FDA/Center for Drug Evaluation and Research, *180-Day Generic Drug Exclusivity* (2001), *available at* http://www.fda.gov/cder/about/smallbiz/generic_exclusivity.htm#COURT ("This 180-day exclusivity provision has been the subject of considerable litigation and administrative review in recent years…").

[282] *See How Pay-for-Delay Settlements Make Consumers and the Federal Government Pay More for Much Needed Drugs: Hearing Before the Subcomm. on Commerce, Trade, and Consumer Protection of the H. Comm. on Energy and Commerce*, 111th Cong. (2009) (Prepared Statement of the Federal Trade Commission), *available at* http://www.ftc.gov/os/2009/03/P859910payfordelay.pdf.

[283] For an example of how exclusivity periods can be extended, *see* discussion *supra* in Chapter 3 regarding stacking of 30-month stay provisions under Hatch-Waxman.

[284] *See, e.g.*, Momenta Comment (12/22/08) at 7 ("The discovery and understanding of the biology of a pathway often allows for patent protection that not only covers the therapeutic protein or antibody itself, but offers the potential to claim coverage of other therapeutic proteins and antibodies that regulate the biological landscape in which the biologic acts."); *id.* at 7; Pearce at 169; Hospira (Wilkie Farr) Comment (12/22/08) at 2; Leicher at 161-62.

[285] Teva Comment (9/30/08) at 6; *see also* Barr Comment (9/30/08) at 8 ("market exclusivity is necessary to encourage companies to develop generic biologics" and that "generic companies need an incentive to undertake costly and time-consuming patent disputes – disputes needed for pre-patent expiration generic market entry."); GPhA (9/30/08) at 5 (generic marketing exclusivity provides "the incentive needed for generic companies to undertake the considerable risk that comes with navigating intellectual property for the brand product and patent. Biogeneric companies will be very reluctant to invest resources if there is no possibly for recouping the costs that come with patent challenges."); W&S Comment (12/22/08) at 2.

[286] Norman at 197 ("to recognize why someone following on after the trail has already been blazed should need any incentive other than the market in and of itself. The market provides plenty incentives for people to do what reasonable persons do every day"); Zielinksi at 196 (The "market dynamic itself will be sufficient incentive, because fewer entrants, and less price discount, so FOBs can make it up in sales."); *see also Competition in the Pharmaceutical Marketplace: Antitrust Implications of Patent Settlements: Hearing Before the S. Comm. On the Judiciary*, 107th Cong. (2001) (statement of Orin Hatch, Chairman , (S. Comm. On the Judiciary) ("Is it necessary or advisable to retain the 180-day exclusivity period given the enormous financial incentives to challenge patents on blockbuster drugs?").

[287] Eli Lilly Comment (12/22/08) at 5.

[288] Allan at 194, 207.

[289] Miller at 198-99; *see also* Amgen Comment (9/30/08) at 19 ("[T]he generic industry is in a very different place today than it was at the time the Hatch-Waxman pathway for approval of generic drugs was adopted. In 1984, the industry was not yet established and success of the generic business model was uncertain.").

[290] Barkoff at 204-05; *see also* Zielinski at 206; Teva Comment at 5 ("Exclusivity periods should be based on the entirety of a particular regulatory and patent system. The exclusivity periods provided in the EU are not a legitimate model for guiding the U.S. since, for example, price controls are prevalent in the EU, while the U.S. does not impose price controls.").

[291] Barkoff at 205-06.

[292] *See* Amgen Comment (9/30/08) at 19.

[293] Hospira Comment (12/22/08) at 7 ("[t]he R&D investment for a biogeneric is significantly greater and could approach $100 million"); *see also* Berhman Presentation at 13; Momenta Comment (9/30/08) at 3; Schultz at 191-92.

[294] Allan at 194.

[295] Hospira Comment (12/22/08) at 7.

[296] Shultz at 194-95.

[297] Brugger at 74.

[298] Shultz at 201-02; *see also* Barr Comment (9/30/08) at 8 (if there were an exclusivity period that it "not prevent the immediate approval of a non-interchangeable, but comparable, generic biologic product" and that it be awarded "to the first interchangeable product to be approved by FDA, rather than to the company that filed the first application seeking approval of such a product, as happens under Hatch-Waxman"); Mylan Comment at 9 (exclusivity provided only to the first biogeneric would not prevent or delay the FDA's approval of a biosimilar product); Novartis Comment (12/22/08) at 18 ("[t]o date, none of the U.S. legislative proposals for FOBs would grant exclusivity to a non-interchangeable FOB. An interchangeability designation is currently considered the most effective way to introduce head-to-head market-based competition with currently-licensed PHS Act biologics."); Teva Comment (10/8/08) at 8.

[299] BIO Comment (9/30/08) at 24-25 ("patent litigation over one FOB product will not necessarily apply to another FOB product," and the risk of litigation free-riders faced in the generic small-molecule context will be much diminished in an FOB context); *see also* PhRMA Comment (9/30/08) at 21 ("[it] is not clear that regulatory exclusivity would be need to encourage patent challenges under an FOB regulatory pathway.").

[300] Norman at 201.

[301] Amgen Comment (9/30/08) at 24 (expressing concern that "excessive patent litigation spawned by the 180-day exclusivity provision" would increase the cost of producing new treatments and cures); *see also* Schultz at 192.

[302] *See* Grabowski at 39; Heldman at 22-32; Heldman Presentation at 3-9; Lane at 36-37, 40; Urlep at 34; Zielinski at 211; Amgen Comment (9/30/08) at 19 ("It appears from the number of biosimilar products under development in Europe, where no market exclusivity is provided for biosimilar products, that market exclusivity for biosimilars in the United States may not be necessary."); Bernstein Research Comment at 12.

[303] For an example of how exclusivity periods can be extended, *see* discussion in Chapter 3 regarding stacking of 30-month stay provisions under Hatch-Waxman.

[304] 21 U.S.C.A. § 355(b)(1); *see, e.g.*, Teva Pharm. USA, Inc. v. Leavitt, 548 F. 3d 103, 106-07 (D.C. Cir. 2008); Ranbaxy Labs. Ltd. v. Leavitt, 469 F.3d 120, 124 (D.C. Cir. 2006); Sandoz, Inc. v. F.D.A., 439 F. Supp. 2d 26, 30 (D.C. Cir. 2006); Apotex, Inc. v. Thompson, 347 F.3d 1335, 1348–49 (Fed. Cir. 2003); Pharma Inc. v. Thompson, 296 F.3d 227, 242–43 (4th Cir. 2002); Andrx Pharm., Inc. v. Biovail Corp., 276 F.3d 1368, 1373-74 (Fed. Cir. 2002); Mylan Pharm., Inc. v. Thompson, 268 F.3d 1323, 1331-32 (Fed. Cir. 2001); Am. Bioscience, Inc. v. Thompson, 269 F.3d 1077, 1080 (D.C. Cir. 2001) (recognizing that the FDA "has refused to become involved in patent listing disputes, accepting at face value the accuracy of NDA holders' patent declarations and following their listing instructions"); *see also Prepared Statement of the Federal Trade Commission Before the Hearing before the S. Comm. on Commerce, Science, and Transportation*, 107th Cong. (2002) (statement of Timothy J. Muris, Chairman, FTC), *available at* http://www.ftc.gov/os/2002/04/ pharmtestimony.htm.

[305] *See, e.g.,* Mova Pharm. Corp. v. Shalala, 140 F.3d 1060, 1065 (D.C. Cir. 1998); Apotex, Inc. v. FDA, No. 06-5105, 2006 U.S. App. LEXIS 10561 (D.C. Cir. Apr. 24, 2006), *aff'd* Apotex, Inc. v. FDA, No. 06-5105, 2006 U.S. App. Lexis 14086 (D.C. Cir. June 6, 2006) ("This case is the latest flare-up in a long running dispute between the Food and Drug Administration (FDA) and several generic manufacturers as to what qualifies under the Hatch-Waxman act as "a decision of a court . . . holding [a challenged] patent to be invalid or not infringed.").

[306] Henry Grabowski, *Follow-on Biologics: Data Exclusivity and the Balance Between Innovation and Competition,* 7 NATURE REVIEWS DRUG DISCOVERY 479 (June 2008) [hereinafter "NATURE"].

[307] Subsequent calculations and adjustments to the Nature model include: Henry Grabowski *et al.*, *Updating Prior Analyses and Responding to Critiques,* DUKE UNIV. DEPT. ECON. WORKING PAPER, No. 2008-10 (Dec. 22, 2008) (hereinafter "Updating Analyses"); Matrix Global Advisors Comment (12/22/08); Alex Brill, *Proper Duration of Data Exclusivity for Generic Biologics: A Critique*, MATRIX GLOBAL ADVISORS, LLC, WHITE PAPER (2008).

[308] Such an approach would require, at a minimum, R&D cost information to which the FTC does not have access.

[309] Joseph DiMasi & Henry Grabowski, *The Cost of Biopharmaceutical R&D: Is Biotech Different?*, 28 MANAG. DECIS. ECON. 469 (2007).

[310] The phase probabilities, as well as the development times, are estimated from a separate database of 522 biologic drugs.

[311] Grabowski, NATURE, at 483, Note 6, Box 3.

[312] *Id.* at 483, Note 9, Box 3.

[313] *Id.* at 485.

[314] *Id.*

[315] *Id.* at 486.

[316] *Id.* at 486.

[317] *Id.*

[318] DiMasi, *The Cost of Biopharmaceutical R&D,* at 474.

[319] Grabowski, NATURE, at 486.

[320] *Id.* at 487.

[321] Brill, *Proper Duration of Data Exclusivity.*

[322] *Id.* at 8 (citing DiMasi, *The Cost of Biopharmaceutical R&D*).

[323] *Id.* at 8-9(citing an alternate source of financial data more recent than that used in the Nature model).

[324] *Id.* at 10.

[325] *Id.*

[326] Congressional Budget Office ("CBO"), Cost Estimate (S.1695), *Biologics Price Competition and Innovation Act of 2007 S. 1695, As Ordered Reported by the S. Comm. on Health, Education, Labor, and Pensions on June 27, 2007* (June 25, 2008), *available at* http://www.cbo.gov/ftpdocs/94xx/doc9496/s1695.pdf.

[327] Zuckerman Spaeder Comment (12/22/08) at 10.

[328] Grabowski, *Updating Analyses*, at 13.

[329] Matrix Global Advisors Comment (12/22/08).

[330] American Enterprise Institute Comment (12/10/08) at 6.

[331] DiMasi, *The Cost of Biopharmaceutical R&D*.

[332] The FTC was unable to determine how the Nature model factors in sales of biologic products outside the United States, including potential sales of biologic products in European markets prior to their approval in the U.S., and European market revenues for the pioneer's branded product after biosimilar entry. Accordingly, there could be additional weaknesses in the Nature model concerning its treatment of international revenues.

[333] Grabowski, NATURE, at 483, Box 3, Note 5.

[334] *Id.* at 485.

[335] Duke University Comment (12/22/08) at 5 ("Specifically, in the Nature article I assume that, starting ten years following launch of the innovator biologic, revenues will begin to decline due to obsolescence at a rate of

3.5% per year. The introduction of new branded biologics by competitors (branded competition with other "first generation" and "second generation" products) is a likely source of this obsolescence.).

[336] *See e.g.*, Office of Technology Assessment, United States, Congress Edition, *Pharmaceutical R&D: Costs: Risks and Rewards*, (1994) at 66.

[337] Matrix Global Advisors Comment (12/22/08) at 3 ("Given the desire to impose conservative assumptions, the Brill model assumes the price decline of innovator drugs is equal the FOB price.").

[338] Dennis W. Carlton and Jeffrey M. Perloff,, MODERN INDUSTRIAL ORGANIZATION, at 93 (4th ed. 2005).

[339] This is likely a conservative assumption as the branded manufacturer may be able to price above the FOB and any ability to do so likely would allow the branded manufacturer to break-even sooner.

[340] *See, e.g.*, Alexis Ahlstrom *et al.*, *Modeling Federal Cost Savings of Follow-On Biologics*, Avalere Health LLC (Mar. 2007) *available at* http://www.avalerehealth.net/research/ docs/Follow_on_Biologic_ Modeling_ Framework.pdf.

[341] Grabowski, *Updating Prior Analyses.*

[342] 21 U.S.C.A. § 301 *et seq.* (2009).

[343] A full NDA is also known as a 505(b)(1) application, referring to FD&C Act § 505(b)(1). 21 U.S.C.A. § 355(b)(1) (2009).

[344] Although biologics must be approved through the BLA process, which center at the FDA performs the review is more complicated. CBER regulates allergenic extracts (*e.g.*, for allergy shots and tests), blood and blood components, gene therapy products, devices and test kits, human tissue and cellular products used in transplantation, and vaccines. The FDA transferred review of all recombinant proteins and monoclonal antibodies, except for hormones such as human growth hormone and insulin, to CBER. Then in 2003, the FDA transferred certain therapeutic biologic products from CBER to CDER. FDA, *Transfer of Therapeutic Products to the Center for Drug Evaluation and Research available at* http://www.fda.gov/cber /transfer/transfer.htm. Accordingly, CDER regulates monoclonal antibodies designed as targeted therapies in cancer and other diseases, cytokines (types of proteins involved in immune response), growth factors (proteins that affect the growth of a cell) , enzymes (types of proteins that speed up biochemical reactions), such as thrombolytics (used to dissolve blood clots), immunomodulators (agents that affect immune response). Additionally cell therapies and gene therapies are reviewed by the FDA's Cellular, Tissue and Gene Therapies. 42 U.S.C.A. § 262 (2009).

[345] FDA's broad regulatory authority over biologic issues, including approval of biologic drug products resides in the PHS Act. The PHS Act also provides the FDA with the authority to: (a) protect the public against threats of emerging infectious diseases, (b) to promote the safe and appropriate use of biological products, (c) inspect manufacturing facilities of biologics before product approval is granted, and thereafter, on a regular basis, (d) monitor the safety of biological products after they are marketed (e) suspend biologic licenses where there exists a danger to public health, (f) prepare or procure products in the event of shortages and critical public health needs, and (g) prevent the introduction or spread of communicable diseases within the country. FDA, *Frequently Asked Questions About Therapeutic Biological Products*, *available at* http://www.fda.gov/cder/ biologics/qa.htm.

[346] FD&C Act defines "drug" as "(B) articles intended for use in the diagnosis, cure, mitigation, treatment or prevention of disease in man or animals; (C) articles (other than food) intended to affect the structure or any function of the body of man or other animals; and (D) articles intended for use as a component of any article specified in clause (A), (B), or (C)." 21 U.S.C.A. § 321(g)(1) (2009). For historic reasons, recombinant human insulin and recombinant human growth hormone ("HGH") were approved under section 505 of the FD&C Act, not under the PHS Act. *See* FDA, *FDA 101: Biological Products*, *available at* http://www.fda.gov/-consumer/updates-/biologics062608.html; *see generally*, FDA's Center for Biologic Drug Evaluation ("CBER") webpage, *available at* http://www.fda.gov/cber/about.htm.

[347] *See* FDA's Center for Drug Evaluation and Research ("CDER") webpage, *available at* http://www.fda.gov/cder/biologics/qa.htm; *see also* FDA, *The New Drug Development Process: Steps from Test Tube to New Drug Application Review*, *available at* http://www.fda.gov/cder/handbook/develop.htm.

[348] Behrman at 17, 19 ("the [FDA's] review of any application, be it drug, be it a biological product, makes an assessment of what is in the best interest of the public given the available information. There will always be uncertainty. There is uncertainty about the simplest small molecule drugs."); *see id* ("Although medical products are required to be safe, safety does not mean zero risk, since all medical products are associated with some level of risk. A safe biological product is one that has reasonable risks, given the patient's condition, the magnitude of the benefit expected, and the alternatives available. The choice to use a biological product involves balancing the benefits to be gained with the potential risks.").

[349] *See* 42 U.S.C.A. § 262; 21 U.S.C.A. § 321 *et seq.*; and 21 C.F.R. 601.2.

[350] *See e.g.,* Ernst Berndt, *et al*, *Opportunities for Improving the Drug Development Process: Results from a Survey of Industry and the FDA*, NBER Working Paper No. W11425, (June 2005), *available at* http://ssrn.com/abstract=745818.

[351] *See* 21 C.F.R. §312.21(a) (2008).

[352] *Id.* §312.21(b).

[353] *Id.* §312.21(c).

[354] *See* H.R. Rep. 98-857(I), 1984 U.S.C.C.A.N. 2647; Letter from Janet Woodcock, Director, CDER, FDA to Petitioners (October 14, 2003) at 6, *available at* http://www.fda.gov/ohrms/DOCKETS/dailys/03/oct03/102403/03p-0408-pdn0001.pdf [hereinafter "FDA's First Response to Omnitrope CPs"]. Generic drugs applications of drugs approved pre-1962 were approved pursuant to the Drug Efficacy Study ("DESI") program upon a showing that they were duplicates of the reference drug. The DESI system was obviated by Hatch-Waxman provisions.

[355] Post-1962 approved drugs whose patents had expired and were available for generic manufacturers, included five best selling drugs: Valium, Motrin, Inderal, Dyazide, and Lasix. *See* H.R. Rep. 98-857(I), 1984 U.S.C.C.A.N. 2647 at 2650; FDA's First Response to Omnitrope CPs at 6.

[356] 21 U.S.C.A. § 355(j)(2)(A)(ii) – (iv) (2009).

[357] FDA's First Response to Omnitrope CPs at 2 (citing Eli Lilly and Co. v. Medtronic, Inc., 496 U.S. 661 (1990), Bristol-Myers Squibb Co. v. Royce Labs, Inc., 69 F.3d 1130, 1132-34 (Fed. Cir. 1995)).

[358] *Id.* at 4, citing approximately 80 drug approvals via the 505(b)(2) process.

[359] *Id.* at 3-4; H. REP. 98-857 (1984), *as reprinted in* 1984 U.S.C.C.A.N. 2647, at 2687 ("The only difference between a NDA and an ANDA is that the generic manufacturer is not required to conduct human clinical trials. FDA considers such retesting to be unnecessary and wasteful because the drug has already been determined to be safe and effective. moreover, such retesting is unethical because it requires that some sick patients take placebos and be denied treatment known to be effective."); Behrman at 24-25.

[360] The Federal Food, Drug, and Cosmetic Act, 21 U.S.C.A. § 301 *et seq.*, as amended by the Drug Price Competition and Patent Term Restoration Act of 1984 (the "Hatch-Waxman Act") and the Medicare Prescription Drug, Improvement, and Modernization Act of 2003, 21 U.S.C.A. § 355(j) and 35 U.S.C.A. § 271(e).

[361] 21 U.S.C.A. § 355(j) (2009). *See* 21 U.S.C.A. § 355(j)(2)(A)(ii)-(iv) (2009).

[362] FDA's First Response to Omnitrope CPs at 2.

[363] 21 U.S.C.A. § 355(j)(2)(A)(ii) – (iv) (2009).

[364] *See* FTC GENERIC DRUG STUDY at 7. The "Bolar Amendment" passed as part of the Hatch-Waxman, reversed the Federal Circuit's decision in Roche Products, Inc. v. Bolar Pharmaceutical Co., 733 F.2d 858 (Fed.Cir.1984), and provided that "[I]t shall not be an act of infringement to make, use, offer to sell, or sell within the United States or import into the United States a patented invention...solely for uses reasonably related to the development and submission of information [to support a market approval to the FDA]"; 35 U.S.C.A. § 271(e)(1); Merck v. Integra, 545 U.S. 193 (2005)(Section 271(e)(1) provides a wide berth for use of patented drugs in activities related to federal regulatory process, including uses reasonably related to the development and submission of any information to the FDA); Eli Lilly & Co. v. Medtronic, Inc., 496 U.S. 661 (1990) (It is not an act of infringement to use or import into the United States patented invention solely for uses reasonably related to the development and submission of information to the FDA for product approvals).

[365] *See* H. REP. 98-857 (1984), *as reprinted in* 1984 U.S.C.C.A.N. 2647 at 2687; PhRMA, *Delivering on the Promise of Pharmaceutical Innovation: The Need to Maintain Strong and Predictable Intellectual Property Rights,"* (April 22, 2002); FTC Generic Drug Study at 7.

[366] *See* 35 U.S.C.A. § 156. See also FDA CDER, *Small Business Assistance: Frequently Asked Questions on the Patent Term Restoration Program, available at* http://www.fda.gov/Drugs/Development ApprovalProcess/SmallBusinessAssistance/ucm069959.htm.

[367] FDA's First Response to Omnitrope CPs at 6.

[368] Section 505 of the Hatch-Waxman Amendments requires an NDA applicant, including some 505(b)(2) applicants, to submit to the FDA (for publication in the Approved Drug Products with Therapeutic Equivalence Evaluations ("the Orange Book") the identifying all US patents that claim the drug substance, methods of formulating, composition of matter, and of method of using the drug and which could be infringed.

[369] Both 505(b)(2) and 505(j) applicants must certify to each reference listed patents when they file their drug applications, stating either that: (1) under Paragraph I that such patent information has not been filed; or (2) under Paragraph II that such patent has expired; or (3) under Paragraph III the date on which such patent will expire, or (4) under Paragraph IV such patent is invalid or will not be infringed by the new drug . No ANDA or 505(b)(2) NDA will be approved by the FDA until all the listed Orange Book patents on the reference drug have expired, or have been successfully challenged by an applicant, or any applicable 30-month stay has expired. 21 C.F.R. 314.107; 21 U.S.C.A. § 355(j)(7)(A); 21 U.S.C.A. § 355(j)(2)(A)(vii); 21 U.S.C.A. § 355(j)(2)(A)(vii)(IV); *see also* H. REP. 98-857 (1984), *as reprinted in* 1984 U.S.C.C.A.N. 2647 at 2655 ("the committee recognizes that in some instances an applicant will have to make multiple certifications with respect to product or controlling use patents. For example, if the product patent has expired and a valid controlling use patent will not expire for three years, then the applicant must certify that one patent has expired and the other will expire in three years.").

[370] These are often referred to as Paragraph I and II certifications. 21 U.S.C.A. § 355(j)(5)(B)(i)-(ii); 21 U.S.C.A. § 355(b)(2)(A)(i)-(ii).

[371] 21 U.S.C.A. § 355 (j)(5)(B)(iii).

[372] 21 U.S.C.A. § 355(j)(2)(B).

[373] 21 U.S.C.A. § 355(j)(5)(B)(iii).

[374] A "new chemical entity" or "NCE" is a drug that contains no active ingredient (including any ester or salt thereof) previously approved under section 505(b); 21 U.S.C.A. § 355, FD&C Act § 505 (c)(3)(D)(ii)-(iv), § 505(j)(5)(D)(ii)-(iv). The 5 year exclusivity provision of the Hatch-Waxman Act applies only to drug products approved under section 505(b) of the FD&C Act and not biologics. *See* FD&C Act. § 505(c)(3)(E)(ii).

[375] 21 C.F.R. § 314.08(a) (2008). An active moiety is "the molecule or ion, excluding those appended portions of the molecule that cause the drug to be an ester, salt (including a salt with hydrogen or coordination bonds), or other noncovalent derivative (such as a complex, chelate, or clathrate) of the molecule responsible for the physiological or pharmacological action of the drug substance." *Id.* CDER makes NCE exclusivity determinations on all relevant applications. CDER reviews all relevant applications, with or without a request from the applicant, for an exclusivity determination. There is no requirement to apply.

[376] 21 C.F.R. § 314.108(b)(4)-(5).

[377] Unlike the five-year exclusivity for NCE, which bars submission of an application, the three-year exclusivity bars approval of an application, so that the agency can accept an application and review it during this time period. Like NCE exclusivity, new clinical investigation exclusivity will not bar approval of a full NDA where the applicant has done the work to support the same change for a drug product.

[378] 21 U.S.C.A. § 360aa-dd.

[379] 21 U.S.C.A. § 355a. The FDA grant of 6 months exclusivity is added to any existing marketing exclusivity or patent protection. This exclusivity incentivizes firms to conduct pediatric drug studies.

[380] There are provisions for a generic company to forfeit the exclusivity period, which occur in limited circumstances.

[381] 21 U.S.C.A. §355(c)(3)(E)(iii); 21 U.S.C.A. §355(j)(5)(F)(iii).

[382] 21 U.S.C.A. § 355(j)(5)(F).

Chapter Sources

The following chapters have been previously published:

Chapter 1 – This is an edited, excerpted and augmented edition of a United States Congressional Research Service publication, Report Order Code RL34005, dated April 26, 2010.

Chapter 2 – This is an edited, excerpted and augmented edition of a United States Congressional Research Service publication, Report Order Code R41270, dated May 25, 2010.

Chapter 3 – This is an edited, excerpted and augmented edition of a United States Congressional Research Service publication, Report Order Code RL33091, dated January 6, 2010.

Chapter 4 – This is an edited, excerpted and augmented edition of a United States Federal Trade Comission publication, dated June 2009.

Index

G

H

I

J

K

L

M

W

Y